BECAUSE I COME FROM A CRAZY FAMILY

BECAUSE I COME FROM A CRAZY FAMILY

THE MAKING OF A PSYCHIATRIST

EDWARD M. HALLOWELL

BLOOMSBURY PUBLISHING

NEW YORK · LONDON · OXFORD · NEW DELHI · SYDNEY

BLOOMSBURY PUBLISHING
Bloomsbury Publishing Inc.
1385 Broadway, New York, NY 10018, USA

BLOOMSBURY, BLOOMSBURY PUBLISHING, and the Diana logo are
trademarks of Bloomsbury Publishing Plc

First published in the United States 2018

Names, locations, and identifying characteristics have been changed to protect the
privacy of the individuals portrayed in this book. Patients and stories are composites.

A portion of the author's proceeds of this book will go to NAMI (National Alliance
on Mental Illness).

Bloomsbury Publishing Plc does not have any control over, or responsibility for, any
third-party websites referred to or in this book. All internet addresses given in this
book were correct at the time of going to press. The author and publisher regret any
inconvenience caused if addresses have changed or sites have ceased to exist, but can
accept no responsibility for any such changes.

ISBN: HB: 978-1-63286-858-9; eBook: 978-1-63286-860-2

Library of Congress Cataloging-in-Publication Data is available.

2 4 6 8 10 9 7 5 3 1

Typeset by Westchester Publishing Services
Printed and bound in the U.S.A. by Berryville Graphics Inc., Berryville, Virginia

To find out more about our authors and books visit www.bloomsbury.com and sign
up for our newsletters.

Bloomsbury books may be purchased for business or promotional use.
For information on bulk purchases please contact Macmillan Corporate and
Premium Sales Department at specialmarkets@macmillan.com.

For Sue

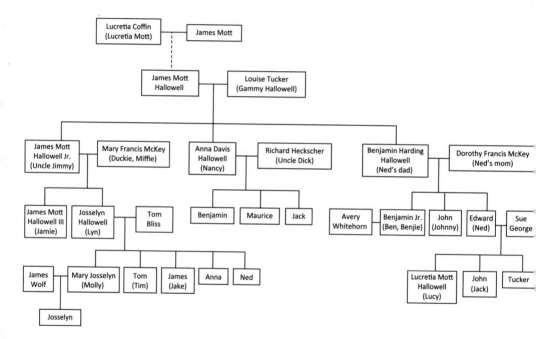

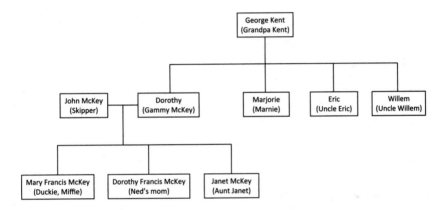

1.

I come from an old New England WASP family, characterized by what I call the WASP triad: alcoholism, mental illness, and politeness. You could be tipsy, even quite sloshed; you could be a bit off, even mad as a hatter; but none of that really mattered as long as you were polite.

The point was never to let life rock you overmuch. Be debonair under duress, be cool under attack, be a good egg. No tears. We specialized in a pitiless pragmatism that deplored sentimentality and revered character. You paddled your own canoe, and if you fell out, well, fare thee well. These things happen. No matter what, we carried on. Rather than ever show sadness, we bucked up. Rather than get angry, we practiced the velvet art of courteous cruelty. You could be as nasty as you wanted to be, as long as you did it with wit and a smile. Above all, your job was to be a good sport. Expressing painful feelings was self-indulgent and embarrassing and created an uncomfortable mess no one wanted to be part of or to clean up. We all knew what a raw deal life could be, but our way of making do was to look it in the eye and, with a tip of the hat, walk on by. If we couldn't beat the devil, we could at least refuse to let him shut us down. Happiness lay in never taking *anything* too seriously. These are my people, and I love them.

But I took a different turn. When I was eleven years old, a voice out of nowhere told me I should become, of all things, a shrink. That was definitely not in the game plan I'd inherited. My people would deem it fine to be a doctor, say a brain surgeon or a cardiologist, but a *psychiatrist*? Please.

Yet there I was, standing by myself on a hot summer day, when an alien voice popped into my head and stated as clearly as a church bell, "*You* should become a *psychiatrist.*" Not knowing what on earth to make of it, I did what I did with most things I didn't understand. I put it aside and moved on.

But eighteen years later, here I was, about to do what that voice from God-knows-where had told me to do. It was the final day of my internship in Medicine. Psychiatrists are required to do a year of a medical internship before beginning psychiatric training, both because many medical conditions can cause what appear to be psychiatric problems and also—for me this was far more important—because the year of internship bonds you to the medical profession and makes you feel like a real doctor, the way I am told boot camp bonds recruits to the Corps and makes them feel like real Marines. Only our medical internship was twelve months instead of a measly thirteen weeks.

Minutes short of being done, I'd finished writing the final progress notes on my patients and was staring off into space, drumming my chewed-up Bic pen on the Formica counter in the nurses' station. It had been years since I'd thought about that inexplicable voice from age eleven, but at that moment in the nurses' station I flashed back to it and laughed out loud. *Little boy has auditory hallucination telling him he should become a shrink . . . and then becomes one.* Not jolly likely, as my Gammy Hallowell would have said. But there it was.

One of the nurses nearby asked, "Did someone say something funny that I missed?"

"No, Nan, don't worry, you didn't miss a thing. God forbid you should miss something!"

"Get outta here, Hallowell. You're finished today, right?"

"Yup," I said. "Thanks for the memories. I'll never forget *you,* that's for sure."

"No doubt you say that to all the girls, but thanks. We'll miss you. You're a good doctor."

Nan had no idea how much her words meant to me. All year I had done my best to keep up with all the brainiac interns who were going into internal medicine. I didn't want to be the weak link headed into psychiatry. Because the nurses were really our best judges, what Nan said capped my year. "Thanks. You guys taught me a lot."

As to why I laughed, that was too much to tell Nan. But here's the story about the voice. I was standing in the shade of some scrawny pines along a dirt road on a sweltering day in July waiting for my cousins to come outside so we could get relief from the heat by going for a swim in the lake below. I can still see my hand resting on the top rail of one of the splintery, weathered split rail fences so common on Cape Cod when a unique voice, unlike anything I'd ever heard before or since, popped into my brain and told me, as if delivering a message from beyond, to become a psychiatrist. Back then, "psychiatrist" was a word I'd never even used and only vaguely understood.

Wearing just my tattered bathing suit, with a threadbare terry-cloth towel over my shoulder, I was alone outside my aunt Janet's house in Chatham, the small town where I lived most of my early years, when that weird voice broke in.

To make the moment even more bizarre, I reacted as if it were *not* bizarre. Instead, not missing a beat, not even doing the logical thing and looking around to see if a real person might be standing nearby, I simply took the message in stride, as if hearing words popping into your brain out of nowhere was a run-of-the-mill occurrence rather than the abnormal event it actually is: a cardinal sign of psychosis.

Even as I took the voice at face value, I didn't get right on it. I actually forgot about it and went swimming. Nor did I determine then and there to become a psychiatrist and pursue that goal the way some kids from an early age single-mindedly work at becoming a professional basketball player or a brain surgeon. The advice the voice gave me got buried.

Still, the voice must have planted some kind of powerful Jack-and-the-Beanstalk seed, because, improbably, here I was, at age twenty-nine, having hoisted my way up the slippery stalk, branch by elusive branch, about to start my psychiatric training.

All kinds of life had happened to me in the interim. I had most definitely not walked the typical path that leads to doctoring—of any kind. There'd been so much chaos in my childhood—insanity, drinking, divorces, violence, sudden uprootings and moves—and there'd been so little planning and guidance until very late that it was, well, "not jolly likely" for me to be standing in the nurses' station at a VA hospital having just finished twelve months of medical internship.

But now I had earned the chance to make good on what that voice had told me to do, and to satisfy my long-standing curiosity about the mind, which had been ignited as a kid talking about people at dinner with my gossipy family.

I knew just from what I'd learned in medical school that psychiatry was nowhere remotely close to having it all figured out, but at least now I could join the search as a certified player. I could learn what others had done and see what I could do myself.

Until now, my main instructors on human nature had been my family and my teachers, as well as Shakespeare, Dostoyevsky, Samuel Johnson, and all the other writers I'd come to love. But now, with medical training, I could also use science as my source, combined with the lives of real people, to take on the complexity of the mind face-to-face.

2.

Looking back on it now, I see the pitch pines and scrub oaks common on Cape Cod, the dogged, stubby trees that greened the scenes of my early childhood, sucking all that they could out of the spare and sandy soil, unashamed of their short stature compared to grander trees off the Cape. Offering their branches to all comers, chickadees, tree sparrows, bluebirds, tufted titmice, the occasional red-tailed hawk, and crows, these spunky little trees stood above clusters of busy bushes that bustled in the breeze, surrounded by tall beach grasses that bent to the wind as one, interspersed everywhere by hardy Cape wildflowers: goldenrod and sunburst, chicory, daisies, daylilies, Queen Anne's lace, and, of course, cattails waving like fat Churchill cigars on thin brown stems near the many ponds and swamps dotting the Cape.

I see a stray snapping turtle, maybe a female on a nesting foray, inching her way from a lily pond near the road, and I see swatches of runaway sand seeping out here and there onto the highways looking like spilled café au lait, giving out-of-towners coming down Route 6 an early tease of the beach, ocean, and bay, never far away.

Once I cross the Sagamore Bridge, arching high over the Cape Cod Canal, I invariably smell the brackish ocean air and feel its slight but welcome nip, while a breeze presents the Cape's unique bouquet of honeysuckle, saltwater, roses, dead fish, and dried seaweed.

As you near the shore, whichever shore you seek, you will see a strip of beach grass guarding the way, whose bite on your feet you can feel as soon as you see it if you've ever walked through beach grass barefoot before—as I did thousands of times as a child—set off along the road by a pointillist array of rose hips, huckleberries, purple thistle, beach plums, and blue hydrangeas.

If the shore pulls you in closer, you will see the dry, tufted portion of the beach with sand that's laborious to walk through, especially if you're carrying a picnic basket, spanning out in front of the band of beach grass. Beyond that is a line of greenish-black seaweed blistering in the sun, marking the last high tide, perhaps concealing a lost horseshoe crab or two deep enough still to be damp. Beyond that lies wet, firm sand creating, at low tide, a vast, glassy surface that scores of sandpipers busily scamper across like nature's committed commuters, their trident footprints and the tiny eddies they stir quickly smoothed over by the sheen of water, while scavenging seagulls, the unofficial mascots of Cape Cod, glide gracefully above, pellet-sized eyes searching, orange tweezer beaks waiting for the snatch, when suddenly, abruptly one of them breaks the serenity of the scene and dives straight down for a fish, a crab, periwinkles, or picnickers' forgotten food.

Craggy wooden lobster pots, dried-up barnacles still stuck to them, stacked up in front yards, decked out with multicolored buoys, often next to an old rust-stained boat in serious need of a paint job up on stilts; makeshift roadside stands selling homemade crab apple, beach plum, or honeysuckle jelly in Mason jars with red-bordered handwritten labels gummed onto them, next to flimsy bushel baskets made of sweetgum slats and wire overfilled with ears of corn, silk hanging from them like brownish-yellow bangs in need of a trim; each town's ballfield spiffed up and festooned with pennants and balloons for the Cape League baseball season; bandstands, town greens, bars with "surf," "clam," "stormy," or "tide" in their names; hunky or paunchy cops whose private lives the locals knew all too well and would pick over as if they were blue claw crabs offered up as a succulent snack; and white churches with sloping lawns next to hardware stores and banks, bars next to gas stations behind which dirt roads curl, bump after bump, downward away from the main drag toward a beach or a lake or a pond—all this and more combined to create the colorful backdrop for many story lines, most unknown to me, that wove around my family and childhood: vines of malice and thickets of love.

The Cape in the 1950s was home to many families like mine—not that there were any families quite like mine. Ragtag misfits, stubborn iconoclasts, beach bums, disillusioned dreamers, skeptical believers, and multigenerational natives whose people had never known anything but life on this sea-beaten, bent-arm-shaped peninsula, the year-round Cape population was full of odd lots, people who loved the Cape primarily because it was set apart from the mainland. Most true Cape Codders valued the canal far more than the bridge over it.

Of course, as a kid, I was just a little boy wolfing down life every day as it was fed to me, looking forward to tomorrow, unaware and unconcerned if my life was like anyone else's or not.

3.

Every day I move back and forth in time, memory continually stitching in the myriad bits and scraps from days gone by that knit the past and the present into a single, ever-growing robe.

But even with memory's help, I can't display the garment whole. I have to pick and choose what to put on display, much as I did as a kid when I'd go with my cousin Jamie—four years older than me and my best friend—to one of the many fields around us in Chatham and select among the array of wildflowers which ones to pluck and bring home to put on the dining room table. My life has been decked with wildflowers of all sorts: uncultivated, lovely outliers, eccentrics, many of whom I loved and learned from.

As I stood in the nurses' station at age twenty-nine, having wrapped up the medical business of the morning, the unlikelihood of my being there enveloped me like a waking dream, as if I were that day entering into a romance come true. That fit; my family was a family of romantics, and we paid the steep price romantics usually do.

Right then I thought of Marnie, one of those wildflowers I grew up with. Perhaps because she was not a romantic at all but rather a practical, opinionated, and brazenly iconoclastic woman, Great-aunt Marnie popped into my mind as I was finishing up my year. "Look at me, Marnie!" I wanted to shout. "It actually happened!"

"Isn't that nice, petty," she would have said, using her peculiar term of affection. "Send me a postal card and tell me all about it." Not a postcard, but a postal card, that was the term she would have used. Certainly do not spend the extra two cents to send a letter when a postal card will do. There's never all that much to say, anyway, now is there? She wasn't cheap, just respectful of every penny because she had to make do on very little.

In our years of boarding schools and college, Jamie, Lyndie— his older sister by two years—and I often visited Marnie at her Boston apartment at 92 Revere Street on Beacon Hill. Marnie's living area was quite bare, as she couldn't afford much, but it felt to us like an enchanted palace, an escape from school and a haven in the big city. We would always bring her a rotisserie chicken from Schrafft's down the street to thank her for having us. She stretched that one chicken to last a week, ending it up in a soup.

She slept on a day bed that doubled as a sofa in her sparse living room. Propped up by remnant store throw pillows, she would lie next to the window listening to the Pops—how she loved Arthur Fiedler and the Boston Pops, which she would conduct in the air with one bony arm as she lay on the bed—and listening to talk shows on the radio late into the night. She loved political talk and controversy, and Jerry Williams supplied both.

She rented out three of the rooms in the apartment, providing what little income she made. She was pretty much a socialist, so the rents she charged were well below market rate. But she wanted to rent to deserving down-and-outers. A hard worker herself, she washed her tenants' sheets in her sink, wrung them out by hand as best she could, then lugged them up several flights to the roof to dry on a line in the open air.

Now and then her brother Eric would appear when he was depressed, taking up temporary residence in one of her rooms. This was long before "depressed" had achieved anything like its current status as a well-recognized medical diagnosis. To most people, including your average doctor, depressed meant at best mentally ill— at worst, weak, lazy, manipulative, parasitic, or possessed.

Marnie didn't see it that way. She knew Eric couldn't help how he felt. He deserved no blame. He needed help, so she took him in. Others would call her a born sucker. We young folk saw her as a kind of rooming-house Mary Poppins, and we loved her.

Eric would retreat into his room like a mole into a hole, turn off the lights, pull down the shades, and stay for however many weeks it took until he was ready to emerge. A couple of times a day Marnie would leave a plate of food on the floor. When hungry, he'd open his door a crack, reach one arm out, and snatch the plate back into his room. When asked, Marnie would casually say, "Oh, don't worry about Eric, he's just having one of his spells." When his depression lifted, he'd don his dark blue business suit, white shirt, and tie and walk down the three flights of stairs and back into the world.

He could also flip out. The McCarthy hearings in 1954 made him so angry he became manic, traipsing over the Boston Common with FBI agents keeping an eye on him because he'd made subversive public statements. He was never arrested, though, nor, thanks to Marnie, was he ever homeless or committed to an asylum, to use the vernacular of the day.

Each time he became manic, Eric would find a new wife, divorcing the previous one. Over his lifetime he married seven women, with whom he had many children. As he was never able to find the right treatment, his moods cycled over and over again. He used Marnie's apartment as his safe haven during his depressive periods and by the grace of God stayed out of serious trouble when he was manic. During his stable periods he was able to work in sales, earn a decent living, and support whichever family he was with at the time. I never knew him, only meeting him for a few seconds one time when he was holed up at Marnie's apartment, but I loved how Marnie so faithfully took care of him. Without her knowing it or meaning to, she became a role model for me.

You could say that Marnie, born Marjorie, was the archetypal wildflower. She was the daughter of a Unitarian minister, my eccentric inventor and man-of-the-cloth great-grandfather George Kent, who'd migrated to Chatham from his home in England. Marnie's sister, Dorothy, was my aunts' and my mother's mother. One of her brothers, Willem, whom I never met, had lived a happy, normal life, I was told, leaving the other brother, Eric, to hopscotch his way through life, producing a flock of children and somehow earning a living while coping with what is now called bipolar disorder.

While Eric married many times, Dorothy and Marnie, great beauties when they were young, each married only once, and not happily.

Dorothy would marry John McKey and have three daughters the world would deem extraordinarily beautiful: my aunt Mary Francis (called Miffie but whom I called Duckie because when I was born she lived on a farm where there were many ducks), my mother, Dorothy (called Doffie or Dodie), and my aunt Janet. John and my grandmother, Dorothy, separated for a while in 1935 but otherwise stayed together, begrudgingly. Marnie, on the other hand, after marrying Charles and giving birth to Rosamond (Rozzie), kicked Charles out when she realized he couldn't make a living.

Marnie was a character who defied diagnosis, other than having a heart rhythm so unusual that the famed Harvard cardiologist Paul Dudley White brought her into his class every year so his medical students could listen to her irregular heartbeat. As an adult, she showed her Unitarian minister father who was boss by becoming an atheist.

Living alone, she kept the entirety of the pittance she lived on in a coffee can under her mattress, or in the oven, or in a variety of other hiding places that she would change often, I guess out of fear that one of her renters might have cased the joint and steal it.

One night there was a terrible fire in the building so the population of 92 Revere had to evacuate. Marnie escaped in her bathrobe and slippers, but once outside, she remembered what she'd forgotten. Breaking through the fire department's barricade, unflappable as ever, bathrobe billowing behind her, she charged back into the still-blazing building. She climbed three flights of smoky stairs and rescued her money from its hiding place—that night, the oven—before making her way back to safety.

An unsentimental realist, she seasoned family conversations with stark opinions. When we were little and would ask what happens after we die, she'd blithely reply, "Blackness, petty, blackness." I'm sure she meant no harm by saying that, and did not mean to scare us; she was merely stating what she took to be an obvious fact and wanted to disabuse us of foolish or romantic fantasies. She attended nearby King's Chapel often, not for "the nonsense they preach" but for the music and the socializing.

Decades later, at a graveside ceremony for a cousin, with the extended family gathered to pay respects, someone asked where Aunt Marnie had disappeared to. As people looked around, she swooped down from a green hill above us, loping as if about to take flight, long white hair flowing behind her in the wind, black dress blowing up and exposing her pasty white legs, while triumphantly brandishing a stunning, clearly professionally done flower arrangement.

"Marjorie, where did you get those flowers?" her daughter Rozzie angrily whispered when she reached us at the graveside, knowing very well her mother would never have paid for flowers.

"Oh, well," Marnie said in a full voice, not the least embarrassed, "I just lifted them off one of the pretty graves over that hill. Why shouldn't I? After all, no one is going to miss them."

4.

When my dad graduated from Harvard in 1936 everything looked rosy. He and his brother Jimmy were headed toward careers in business, and their exceptional intelligence, charming personalities, excellent upbringing, and fine schooling predicted they would shine. Gammy's middle child, Nancy, constitutionally the happiest and most balanced of the three Hallowell siblings, but not gifted with the extraordinary intelligence her brothers possessed nor cursed with the psychological difficulties each of them wrestled with all their lives, found a good and stalwart man to marry and embarked on the wonderful adult life she did, in fact, live. She and her husband, Dick Heckscher, had three boys, my cousins Ben, Maurice, and Jack, all of whom went on to make happy families themselves as well as to find great professional success. Accomplished athletes as well, Ben and Maurice are the only pair of siblings in the U.S. Squash Hall of Fame.

Gammy lost her much older husband years before her children were teens. A single mother, she presided over the family like the Social Register matriarch she was. That her two sons picked girls whose family had little money didn't matter as much as the fact that they had private school education, beauty, and charm. Gammy embraced them, as well she might have, because both McKey girls, Miffie and Doffie, were gems by any reckoning.

While Gammy had always had money, thanks to her husband, she became wealthy after Jimmy sold the family house in Chestnut Hill during the Depression and invested most of the proceeds in a stock called IBM. The family lived off that income for the next seventy-five years.

Jimmy started out working in finance in Boston, but he discovered that the business-suit life wasn't for him. In retrospect, I can see he had a major anxiety disorder, but back then, in the 1930s, anxiety was not a condition you admitted to, especially if you wanted to maintain a respectable career and a prominent family name. That a large portion of the names in the Social Register suffered from one kind of mental illness or another was ignored, if acknowledged at all.

Rather than getting help for his crippling anxiety, Jimmy, who was literally a fighter, having boxed bantam-weight at Harvard, told Gammy that he didn't like the pretentiousness of Boston finance and high society, and that he wanted instead to be a farmer. No doubt aghast, Gammy, who was above all else loyal to family, agreed to subsidize the cost. Jimmy had made the money, after all, even though he had turned it over to his mother.

He bought a farm in Pepperell, Massachusetts, just west of Boston, named it Cloverluck Farm, and for a time turned it into a successful dairy operation. His wife would have vastly preferred to be married to a socially prominent businessman, but she went along with her husband.

Even though she complained about Jimmy all the time because he wouldn't attend parties or go dancing, and he insisted on drinking beer out of a can, dressing down, wearing sneakers and tattered trousers, having his hair cut as short as possible (Duckie cut his hair in the kitchen), and looking as little like an executive as he possibly could, I always believed that Duckie loved him. You can tell when there's true poison in a person's resentment of another, but I never sensed anything close to that between the two of them. Frustration and superficial resentment for sure. But also what I perceived as love.

Jimmy, teeming with anxiety he did his best to hide, harbored a lifelong terror of death. As he got older, and it became clear after a lifetime of smoking cigarettes, then a pipe, that his lungs were not going to let him live much longer, his mortal fears intensified.

I never saw Duckie console Jimmy on that score, or on any score for that matter. That's not how they interacted around us kids. But I'm certain that as they went to sleep at night, she'd say something like "Dying is no more than just going off to sleep. Catching forty winks. There's nothing to be afraid of. You just say nighty-night and go peacefully off to dreamland."

At Jimmy's grave, Duckie's point proved, she reassured him one last time: "There, that wasn't so hard, was it?"

Uncle Jimmy was not merely a bundle of nerves, he had a unique style and a certain panache. It makes sense to me that a woman as smart and universally appealing as Duckie fell in love with a man some would deem if not a misfit then at least a cranky icono-clast determined to hide from the world. He actually could be very charming, he had a great sense of humor, and Duckie loved to laugh. Jimmy was also an inveterate practical joker. Unlike my father, Jimmy was too short for most varsity sports, and he loved football, so at Harvard he became manager of the football team, carrying on a long-standing tradition of pranking the Yale team the day of the Harvard-Yale game.

In Jimmy's senior year the game was in Cambridge. The night before the game, Jimmy got the security people to let him into the Yale team's locker room. Using his knowledge of electricity, and his innate creativity, he rigged up an apparatus in each of the urinals and toilets that couldn't be seen so that when a person used the urinal or toilet an electrical current would zip up the stream of urine and zap the player where he least wanted to be zapped. Before the game the next day, people wondered what caused all the yelps coming from the Yale locker room. Harvard won, and Uncle Jimmy liked to say he supplied the electricity.

Duckie could be just as playful. Shortly after they were married, they lived in an apartment on T wharf in Boston. Having believed they'd lucked into a fantastic steal of a deal, they soon discovered why their rent was so low. On the floor below was a brothel. Jimmy found out first but kept it to himself. Duckie always wondered why cabdrivers gave her funny looks when she told them her address. When Jimmy finally told her, she burst out laughing. As luck would have it, they soon discovered a knothole in one of the floorboards that gave them a perfect view of the goings-on in the brothel. They'd see all sorts of well-heeled Bostonians coming and going. One night Duckie saw Arthur Fiedler enjoying himself with one of the hookers. Far from disapproving of Fiedler, she thought all the better of him for it.

If there is a gene for humor and practical jokes, almost all of us had it. For example, when Janet, the younger sister of Duckie and my mother, got married, the wedding was held in Pepperell because the family was still living on the farm. Uncle Jimmy and Uncle Willem (Uncle Eric and Aunt Marnie's sane brother) cooked up a plot using my older brother, Ben, who was ten, as the doer of the deed.

Uncle Jimmy found a big old bullfrog on the farm, which he gave to Uncle Willem, who was charged with taking care of Ben during the wedding. Willem put the frog in his pocket, and then he gave the frog to Ben and told him to put it in *his* pocket until the service began. Willem said he'd give Ben a sign when it was time to release the frog.

The pews were full at the Pepperell Congregational Church when the minister started his remarks. When Willem gave Ben the

OK to release the frog, it immediately hopped forward under the next pew. Suddenly the minister's remarks were interspersed with loud croaks. No one knew what to make of it. As Ben and Willem started giggling, there came a woman's muffled scream. After a few more sonorous croaks, someone grabbed the frog and managed to muzzle it. At the reception, Ben, Willem, and Uncle Jimmy were personae non gratae. Janet's mother, Gammy McKey, was fit to be tied, which is how she usually felt about Uncle Jimmy anyway.

The few serious ones in the family, like Aunt Nell and Gammy McKey, often became the brunt of the jokes. Comedy so depends upon the person who will not laugh that there is a term for that person: the *agelast*. Those two were dependable agelasts.

As is so often the case with comedians, we harbored lots of turmoil underneath our jokes and pranks. Uncle Jimmy couldn't stomach working in the financial world, so off the couple went to Pepperell and Cloverluck Farm. According to all that I would later learn, Duckie worked as hard as Jimmy did, organizing the farm-hands, doing all the cooking and cleaning, making sure everyone was healthy and well fed, and getting her hands dirty every day. Now and then she went waltzing with her father, whom we kids called Skipper, at the Rainbow Room in New York City to the music of Guy Lombardo and the Royal Canadians. She and Skipper became friends with Guy Lombardo.

But mostly she worked the farm. They worked so hard that every afternoon they had to take a nap, or "catch forty winks," as they called it. Even as a three-year-old, I grabbed on to that phrase, wondering how you catch winks. Years later I'd learn that Duckie and Jimmy reserved that time for making love.

While Jimmy and Duckie were working the farm, my dad was working at Goldman Sachs in Boston and living with my mother on Newbury Street. They had two healthy baby boys, three and a half years apart, my brothers Ben and John. Life was good.

But then World War II hit. As a farmer, Jimmy was given an exemption, but Dad volunteered for the Navy, leaving my mother and brothers behind. Dad left his promising job in investment banking, his idyllic marriage, and the sons he adored to become the captain of a destroyer escort in the North Atlantic, fighting U-boats. His life, and the life of my family, would never be the same.

Meanwhile, Uncle Jimmy and Duckie kept the farm going. Jimmy made a deal with the government to employ German prisoners of war as farmhands. Jimmy spoke fluent German, and Duckie had learned a little of it on a trip they'd taken together to Germany, so everyone was able to communicate. The POWs all liked working there so much that no one ever tried to escape, which would have been easy to do.

My older brother Benjie told me Jimmy named the first herd of cows after Greek goddesses. Ben, who had to bring these cows in at age six, told me Athena was the nicest and Aphrodite was the biggest pain in the ass to drive into the barn. He said the biggest, baddest, meanest bull ever was Kitch (short for Lord Kitchener), and his sidekick, Beau Brummel, was no day at the beach either.

Jimmy named the next herd of cows after Shakespeare's women: Desdemona, Ophelia, Cleopatra, and my favorite, Mistress Overdone. I can just imagine coming up to this cow as she was chewing some grass and saying, "OK, Mistress Overdone, time for milking."

Tapping into his science background from college (he took a double major in physics and Romance languages, graduating magna cum laude), Jimmy invented a kind of corn that crows didn't like to eat. He did most of the veterinary work himself, delivering calves and diagnosing diseases, calling the real vet only when he couldn't handle the problem himself, which was not often.

After years of trying, on the brink of giving up, Duckie unexpectedly became pregnant, so my mother starting helping out at the farm. She wasn't much of a farmer, though. She had a way of getting out of the scut work. In fact, my mother was more or less allergic to any kind of work that generated an income. She never held a paying job her whole life. She wasn't lazy—she raised three boys—but she was born to take care of others and be taken care of, it seems, rather than to work in the world.

Benjie was six and Johnny two when Josselyn—my cousin Lyndie—was born. Four years later came Jamie, to the utter astonishment of Duckie and Uncle Jim, who never thought they'd have a first child, let alone a second.

When the war ended, I arrived on the scene, four years after Jamie.

The farm got sold after the war because the farming business was down. We all moved to Chatham, a half-hour drive from Wianno and Gammy Hallowell. Uncle Jimmy opened a bowling alley, and Dad got a job at a boatyard named Ryder's Cove. Life was good, at least as far as my three-year-old self could tell. That's why I don't understand why Dad left. It's kind of a haze. The story that they got divorced because Dad had a serious mental illness doesn't add up, considering how in love they reportedly had been, and that Dad was getting better.

Less hazy is the day Cloverluck Farm was sold. That sale sticks in my memory because of the auctioneer. The sound of his voice as he auctioned off our livestock hypnotized me. I'd never heard anyone talk so fast, or talk—almost sing—quite like that. The words were incomprehensible but they totally held my attention. "Gimme one hunnerd, hunnerd, hunnerd, one hunnerd in the back, gimme two, who'll gimme two, how 'bout hunnerd fitty, who'll gimme hunnerd fitty . . ." I sat and listened, transfixed, for what seemed a long, long time.

That auctioneer's amazingly rhythmic jive was like an introduction to rap music. I remember not being able to comprehend why so many grown-ups were standing around with tissues and hankies to their eyes, crying, because I was totally happy, listening to the music of the sale of our home, Cloverluck Farm.

It would be just the first of many unexpected moves.

5.

The night that Dad left, I was three and a half. I remember sitting on the living room floor with him in our little Cape house in Chatham playing matchies. This was a card game he'd made up, a game uncomplicated enough for a little boy like me to learn. As the name implies, the object of the game was to find matching cards, drawing one at a time from a deck. Simple-minded as it was, I never tired of this game. And even though it must have been really boring for Dad, that particular night he played it as long as I wanted him to.

I was actually tired when Dad put me to bed. He told me a bedtime story involving the character he'd made up: Johnny Creep-mouse. My father was a phenomenal storyteller. It was a gift he had, as did many members of my extended family. I was always rapt as I heard the many ways Johnny Creepmouse took down bad guys and made the world safe. I never questioned for a minute how a mere mouse could make drawbridges rise and tall ships sink. I only wish my dad had taped some of these stories. I felt then the way all kids feel listening to stories they love: completely given over to the realm the tale creates, excited but safe as can be, not a worry in the world.

That night, however, my sense of safety in the world shattered, never to return. I would build lean-tos in life but never again feel the undiluted security I felt when playing matchies on the living room rug that night with Dad.

My mother came in late and sat down on my bed. "Neddy," she said, "your dad is moving out tonight, right now in fact. He's going to live with Gammy Hallowell."

"Why is he moving out?" I asked, nonplussed.

"Because your father and I have decided that we should not be married any longer. We are getting a divorce."

I felt a pang of panic. "Will I still see him? Can we play in the shop?"

"Of course," my mother said. "Right now he's gathering up his tools. At Gammy's you can visit him as often as you want to."

"Is there a place for his workshop there?" I asked, needing details.

"Of course. Gammy's house is huge, you know that. I bet he creates a spot inside the garage for his tools where you can play with him."

Hammers and saws floated through my mind, planes and chisels and a vise the size of a melon, as I tried to understand what was going on, which I was unable to do. I'd never heard of divorce, so I didn't comprehend what Mom was telling me. I just knew it was not good.

"Go to sleep now," she said. "Everything will work out for the best. I promise."

Up until then, my mother had never promised without meaning it, so her word meant something to me. Still, I didn't get it. To this day, I don't know for sure why they got divorced. My mother would later declare it was because the doctors said he had an incurable mental illness and that she should divorce him, but the months leading up to that fateful night he was living with us, sane, and holding a job at Ryder's Cove, a local boatyard. All I know is what I knew even then, the night he left: It was a really stupid thing for the two of them to let happen.

For a long time, I would wonder about it. If he was healthy and out of the hospital, why leave us? I knew he loved me and my two brothers. What had happened? Had I done something wrong?

I didn't know.

Nor would I know for the longest time how my mother met the next man to come into her life, what he was up to, and what his game was. I wouldn't ever know for sure why Unger Stiles, whom I at first embraced and loved like a father, would later change so dramatically.

6.

While I was growing up, everyone was too busy being who they were to pay attention to what other people thought. It wasn't out of arrogance or the belief that we were better than them—although in retrospect I imagine some outsiders might have thought so. It was more that we focused on the details—the old milkcan top from the farm that we used for Bessie's dog dish, or the church key Uncle Jimmy used to open his can of Pabst Blue Ribbon—rather than what other people thought about anything.

We weren't loners, though; Duckie saw to that. She was a true social butterfly, a world-class connector. She made her rounds every day, stopping at various houses for a drink (bourbon or scotch on the rocks with water), or what her father, known to us as Skipper, called a "heist," before she came home to cook dinner.

As a kid, no one ever told me to do well, dream big, or aim for success, and that was fine with me. I had no notion of a trajectory into adulthood. I was unaware of any expectations regarding performance in school or anywhere else, other than Skipper giving me a quarter if I got a good report card, which was not difficult to do. As mentioned earlier, the only expectation of me was to be polite.

The lack of pressure or encouragement was not due to neglect. My family was not interested in advancement, status, or appearances the way I guess most families are. Whatever the opposite of social climbers is, we were that. It was a loving, sprawling family but very much in the moment, with little mention of or apparent care about the future.

What mattered, what so much depended on, and defined each day were the common details, like the burnt toast with marmalade served us before school by my never-married, highly literate great-aunt Nell (Skipper's sister, the McKey side of the family, while Marnie and Dorothy were Kents), who was rumored to have had a dalliance with Wells Kerr, a revered dean at Exeter, but otherwise had lived a life free of romance. She'd met Wells Kerr because her brother Skipper had gone to Exeter. I don't know how in the world she managed to get intimate with Kerr, which makes me doubt the rumor, but on the other hand, she was phenomenally literate, as was he. This is how it so often is with family lore—the people who would know all the details for sure die before the whole truth is known.

For some reason, Aunt Nell loved her toast burnt and spread with marmalade. She turned Jamie and me into fans of this treat as well. After eating it, the three of us would count the new morning glories that had appeared since the day before, climbing the trellis next to her kitchen door, before we walked off to school while Aunt Nell shooed away cowbirds from bothering her beloved chickadees. So many details filled the bubble in which I lived, details like the old foghorn I'd hear on my way to sleep, that still rise up in the sun-rimmed scenes I see when I reenter that bubble by looking back on my early years. I carry a perpetual slide show in my mind that varies each time I view it, scenes zipping by, none allowing enough time to be fully viewed, each emotionally striking a nerve, yes, there it is, there I was, there we were, how green it was, how white-capped and blue, how dim the moon on foggy nights, how much fun, how free of what fills my every day today, and yes, that's the lily pond and there's the leather mail bag Uncle Jimmy took with him every day, and there are the old golf clubs propped up in the corner of the vestibule, how lost and gone forever it all is now and yet how close at hand, tantalizing, irretrievable. Olly olly oxen free.

Memories speed up after Dad moved out, when I was living with Jamie's family on Kettle Drum Lane, just off Shore Road, right next to the Chatham Bars Inn. Aunt Nell lived in a separate house across the yard. Uncle Jimmy had bought two houses when we moved from the farm in Pepperell to Chatham, one house for his immediate family and one for Aunt Nell, Duckie's aunt. We were all intertwined, my extended family, which I thought was wonderful.

Later, in my psych training, I would be taught that intertwined families like mine who relied so much on one another to the exclusion of the outside world were called "enmeshed," deemed by professionals not a healthy way to be because such people do not develop true autonomy but depend too much on one another for identity and support, instead of making their (our) own way in the world independent of the family. I thought this way of looking at us, while true in its way, was too harsh. I liked enmeshed just fine. Enmeshed saved my bacon.

I found (and still find) many of the words psychiatrists use overly weighted toward sickness and pathology. They overlook the strengths in the people they purport to describe and in so doing demean us. Some of those words, like "enmeshed," put me on the defensive and make me want to say, "Oh yeah? So you think you could do better?" Enmeshed sure beats lost and lonely. Instead of enmeshed, I go for terms like "interdependent" or "heavily reliant on one another."

At dinner with Jamie's family, we didn't talk about plans or ambitions. We spoke about the family or played ghost, a word game. For quite a while, I spent most of my time at their house. I can't remember why I lived with them or exactly where my mother and my older brothers, Ben and John, were after Dad left.

The object of ghosts was to add a letter to a growing word but not to complete the word. So, for example, when it came to me, if the letters so far were *s, t, a, m,* I would have to add a letter, but if I added *p* I would lose, as that would spell "stamp." So I had to come up with another letter. Let's say I added a *q*. The next player could challenge me, and if I did not have a real word in mind—there are no words I know of with the combination "stamq" in them—then I would lose, and be given a *g*. You lost when your score spelled out "ghost." I could also add a letter to the front of the word. So, for example, I could add an *e* to the front, and if challenged, say the word I had in mind was "prestamp," and we could go to the dictionary, which we often did, to see if "prestamp" was an actual word. Or I could add an *i* at the end, my word in mind being "stamina." Playing this game night after night fueled a lifelong love of words and wordplay.

Sometimes at dinner we talked of distant relatives, a famous admiral, Great-aunt What's-her-name, or Cousin So-and-so. Now and then, an illustrious name would pop up, although at the time I didn't know it was well known—like that of my great-great-great-great-grandmother, Lucretia Mott, or another friend of the family, Oliver Wendell Holmes, whom the grown-ups referred to as Justice Holmes, and the various Hallowells after whom some gates at Harvard were named. These were all just names overheard at the dinner table as a little boy, but they were never brought up reverentially as if they or, by extension, we were anyone special. They were just names in the flow of conversation, names meaning far less to me than Horace and Joe, the barbers at the bottom of Seaview Street who cut Jamie's and my hair in the waxed crew-cut style of the 1950s, or Roy Bearse, the man who owned the grocery store and gambled on the puppies at a dog track off the Cape, or Benny Nick, short for Benny Nickerson, the portly, friendly cop we'd usually see on our way to school, or Mrs. Forge, short for Mrs. Forgeron, the nice woman with a slight limp who'd come to clean our house and do laundry once a week.

There was no big picture of life that I was aware of. It was the everyday moments that counted: the pork chop Duckie fried up along with the potatoes she mashed, and the frozen Birds Eye peas that would land on my blue Canton china plate, which I was supposed to "clean."

As I said, manners mattered. Stand up when a lady enters the room. Always say please and thank you. Shake hands firmly, squeezing hard, and look the other person in the eye. To this day, if someone doesn't look me in the eye when he shakes my hand, my opinion of him drops, and if his handshake is weak, I really want nothing more to do with him.

I was incessantly told "Don't interrupt," so much so that I embarrassed myself by telling friends, at age six, not to interrupt. A pompous six-year-old is not a popular boy. I quickly learned not to tell people my own age not to interrupt.

Politesse is only part of the WASP triad mentioned before, which also includes alcoholism and mental illness.

We're certainly not all alcoholic, but we do seem to love our cocktails. While an Italian, French, or Jewish party will pile up elaborate foods of all kinds and also serve alcohol, a WASP party will flow with alcohol of all kinds while offering Goldfish, peanuts, and cherry tomatoes as hors d'oeuvres.

Mental illnesses also seem to swim in our genes, even as we pretend otherwise. A WASP stiff upper lip is de rigueur, part of our code of honor, but the family trees of most of us include an ample sampling of depression, anxiety, bipolar disorder, schizophrenia, and a smorgasbord of eccentricities bordering on psychoses, often medicated with alcohol, our go-to remedy for whatever ails us.

The drinking was usually not out of control, and when it did get out of control, as with Uncle Jimmy and Johnny, those people were able to stop drinking altogether. Iron will is another attribute

WASPs tend to have. Unfortunately, though, my mother never quit drinking. In her mind, the benefits of alcohol exceeded the costs. More than a few of us reached that conclusion.

The mental illness part of the triad exacted the greatest toll. It cost my parents their marriage. My father, my brother Johnny, my aunt Janet, and Uncle Jimmy lost the careers they could have excelled in. Numerous other relatives and ancestors missed the normal joys of life because of serious mental illness.

However, despite our complicated struggles, I always felt loved. I enjoyed just about every day of my life, until we moved to Charleston.

7.

When I was born, my mom said she held me in her arms in the hospital and looked down at me with high hopes. "You were the cutest baby in the world," she told me. "You had two perfect dimples and a little mole on your cheek. I said it was where the brownie kissed you. I knew you'd make us all happy again."

So those were my marching orders. Make people happy. It seems most of us are born into certain roles, and that was mine. That it ended up being what I would do for a living speaks to the powerful, invisible forces that shape each of us from the moment of conception.

I learned many years later from my mother that I was conceived when my dad was on leave from the mental hospital. Mistakenly given a pass when he shouldn't have been, upon arriving at home he decided, for no apparent reason, that he wanted to murder my mother. There was my mother facing a psychotic man in a murderous rage who happened to be her husband.

But she was amazing at handling men; it was her great talent in life, even though things didn't work out well for her. When my father told my mother he wanted to kill her, she didn't freak out. How she remained calm, I don't know, but somehow she talked him into making love instead of killing her. (She told me this herself.) After they made love, Dad got out of bed and put on his galoshes. After a while my mother heard a loud noise and went outside where she found him standing stark naked in a snowy cornfield—they lived on a farm outside Boston. Wearing nothing but his galoshes, he was shooting crows out of the sky with a shotgun. He thought the crows were Nazis. He'd draw a bead on one, then *bam!* blow one up, then another and another, leaving feathers and blood all over the field. People on the adjoining property called the police, and the police came, but the poor cops didn't dare go anywhere near my father when they got there. I guess disarming a naked man known in the town to be a war hero who's shooting off a shotgun was not something they were trained for. They looked to my mother, who, artful as ever, sweet-talked my dad into giving her the gun and coming inside.

That's how my mother became pregnant with me: a unique beginning for a psychiatrist. But even though I was supposedly born to help others, I was anything but selfless as a little boy. In fact, I was pretty selfish, if the stories I heard about my fourth birthday party are representative. I didn't want to share anything with anyone. But I was cute, innocent, trusting, and eternally naïve, and I had a natural sparkle that brought joy to most places I went, or so I was told years later.

Since when I was born my father was in a mental hospital, my extended family all lived on my uncle's farm to give my mother help with my two older brothers and me. Mom would visit Dad several times a week.

Nobody talked much about my father's condition, as insanity was thought shameful, not a state of mind any respectable person ever entered, as if it were a venereal disease, contracted voluntarily and irresponsibly. If she mentioned him at all, my father's mother, Gammy Hallowell, would tell people her son Ben was in need of a rest.

When I was a toddler, Dad was in different hospitals, Taunton State, Baldpate, and the VA. He was diagnosed as schizophrenic and underwent numerous shock treatments, both electric shock and insulin-induced hypoglycemic shock.

When I was about three, a smart young doctor put Dad on a new medication, lithium, and—presto!—he got better and soon thereafter was discharged. If he'd been put on lithium sooner, my family might have been saved.

8.

After Dad moved out, even though I think Ben and John were at home in Chatham, I didn't see them much. Instead, Jamie, Lyndie, and I became a close-knit trio—enmeshed for sure. We were one another's lifelines.

Still, my lifelong feeling of being an outsider got its start back then. My two actual siblings were so much older than me that I couldn't relate to them as close friends, while Lyndie and Jamie, my actual best friends, were not my true siblings. They almost were—which was great—but I was one step removed.

Jamie and Lyndie had each other growing up, but even they shared a bit of the outsider feeling that I felt. Maybe it was because we were not natives of Chatham, or because anxiety ran so rampant in our genes. Maybe because none of us was particularly popular in school or in town. Whatever the reason, we banded together because we needed a safe haven in a world in which we did not feel we entirely belonged. With one another, we felt at home. It's why the enmeshed thing was such a blessing.

Back then—it was the 1950s, Ike was president and Elvis ruled—Lyndie was the queen. She was six years older than me and two years older than Jamie. Uncle Jimmy doted on her, as his only daughter, and gave her whatever she wanted. She was able to get Jamie and me to do pretty much anything she wanted us to do.

One day, for example—I guess I was ten or so—she wanted a special cheeseburger called a 3-D (the progenitor of the Big Mac) from Howard Johnson's, which was at the other end of town. Lyndie didn't have her driver's license yet, so to get the 3-D someone had to bike there.

"Neddy, would you ride down to Ho-Jo's and get me a 3-D?"

"Why should I? I'm not hungry."

"Well, why don't you do me a favor and get it?"

"Because I'm happy here doing nothing."

"Well, if you're doing nothing, why not do me a favor and go get me a 3-D? Since you're not doing anything anyway."

"Why can't you go get it if you want it?"

"Because I'm reading this magazine. Please, just go get it. It won't take you but a few minutes."

The next thing I knew I was on my bike on my way to the other end of Chatham to fetch Lyndie a 3-D. She never lost this power over me or Jamie, even when we were adults. When she was nice, she was *soooo* nice, but when she was angry or distant it felt so terrible that both Jamie and I did whatever we could to stay in her good graces.

Her hold on us did not come just from her being the oldest. She seemed to have a magical skill to make life fun and exciting. She came up with ideas for things to do. She was the one who made Christmas special, not the grown-ups. And she was the one who always wanted to know what was going on in my life, far more than my own mother did.

Lyndie was pretty and also really bright, which set her apart from the other kids in her high school class. She had two close friends, Nancy Thornton, who was also smart and pretty, and Martha Toabe, one of the few Jewish people in Chatham, daughter of an attorney named Igo, a name we thought hilarious. Martha was extremely shy and introverted, and quite an odd duck, but also intelligent and sensitive. Lyndie always loved and identified with oddballs.

I didn't know what Jewish meant, or Protestant or Catholic for that matter. That Martha Toabe was Jewish meant absolutely nothing to me. I was too little to know of the prejudices in Chatham. I would later learn that there was a local man whose garage was full of Nazi memorabilia. He held meetings there from time to time to vent his hatred of Jews, as well as all other nonwhite groups, not to mention homosexuals, one of his most despised targets.

Unbeknownst to me, Jamie was discovering during those years his own sexual feelings for boys. Just as I didn't know what Jewish was, I also didn't know what homosexual was, but Jamie did. He kept it secret because this was the 1950s in small-town America, where being gay was about the most forbidden thing a person could be, right up there with being a Communist. Looking back, I feel so bad for him, having to hide who he was and feel ashamed.

Jamie knew about the man who loved Nazis and hated gays— fags, fairies, queers—and was terrified of him. He was a large, muscle-bound man who openly spewed hatred wherever he went. Until Jamie told me years later, I had no idea of the daily fear this Neanderthal caused him. The town seemed so sweet and peaceful on the surface. I never sensed the pockets of fear, rage, bigotry, and other all-too-human poisons hidden underneath.

There's so much I just didn't see. I suppose Chatham, with its select vindictive people like the Neanderthal, resembled most small towns. As cruel as the community could be in rare spots, though, it was mostly filled with good and caring people, so many that I never felt afraid. I remember almost all of them with a smile, fondness, and even love, from Mrs. Eldredge, my first grade teacher, and Mrs. Brown, my second grade teacher, whom I would visit by riding my bike to her house for milk and cookies, to Mr. Tileston, who conducted the band concerts every Friday night (they *still* have them, sixty years later), to Ben Bassett, the angular, handsome high school principal who dressed all in black when he played golf, to the town drunk.

As an adult, I mentioned "the town drunk" one day to Jamie. He laughed and said, "*The* town drunk? Who did you have in mind? There were quite a few candidates for that title." It's true, alcohol flowed so freely in Chatham that some people called it a drinking town with a fishing problem. But if it was one, it was a hugely friendly drinking town. Chatham was a happy place to live, or so it seemed to me.

Even though our family owned a local business, the bowling alley, and we kids went to the local school, we did not qualify as true Chathamites. We'd moved from a farm in western Massachusetts, so even though my dad and Uncle Jimmy had spent much of their childhood in Wianno on the Cape, where they were well known for winning many sailing trophies, we had nowhere near native status.

I never knew how much trouble Lyn had fitting in at school until years later when Jamie told me. Duckie and Uncle Jimmy knew she was having trouble socially, so they sent her to the George School, a Quaker boarding school outside Philadelphia, but she got so homesick that she came back and finished up at Chatham High School in spite of social problems there.

While smart and pretty, she was also shy and sensitive, and teased by the boys. One day they fed a turkey Ex-Lax, broke into the car Lyn drove to school, and locked the turkey inside. When she came out to her car, she found a berserk turkey and a car full of turkey shit. Mortified, she somehow had to keep her composure as the boys whooped it up while she drove home in the stinking car, all the while fending off the rampaging bird. Uncle Jimmy would slaughter the turkey himself, and we'd eat it, a kind of revenge.

Still, Lyn didn't want to leave Chatham. In addition to the family, she had a charismatic English teacher who got both Nancy and her excited about literature. He recognized that these two students were exceptional. With his guidance, Lyn's lifelong love of words, stories, books, life's details, and people's quirks, all of which were in her genes, and in the genes of all of us kids, continued to develop.

She'd go with this teacher down to the boatyard to work on his sailboat or, in good weather, go out sailing. She'd go to his house to watch Shakespeare on TV with him and his wife now and then, once bringing Jamie and me. I was bored to death.

We roamed Chatham together: Lyndie, Jamie, and me. At the center of town there was a candy-newspaper-magazine-stationery-some-of-everything store called the Mayflower, where I leafed through my first *Playboy* magazine. Across the street was the corner drugstore with its soda fountain with round stools with padded green seats that spun around plugged into the floor on shiny aluminum columns. I loved spinning on those stools between sips of my frappe.

The proprietor was Mr. Parmenter. He never could get Jamie's name right. He always called him Janie. Jamie desperately tried to correct him a few times but finally gave up. Jamie hated it because he thought others would realize that he liked boys instead of girls, and that Mr. Parmenter was on to him, which is why he called him Janie.

Lyndie would often ask me to go into the drugstore and buy Tampax for her. About eight, I had no idea what Tampax was. But if Lyndie wanted it, I would get it. So I'd just walk right up to the counter at the back of the store, only a few feet away from all the people having their hot dogs in grilled, buttered buns and Cokes or frappes at the soda fountain, and, as if I were asking for a bottle of cough syrup, say, "I want to buy some Tampax." I could never understand why the lady behind the counter would smile and people at the soda fountain would giggle as I purchased the blue and white box of Tampax, which the lady behind the counter would promptly put into a white paper bag.

One day—I was around ten, so Jamie was fourteen and Lyndie sixteen—we stopped at the open-air Main Street fruit stand. As we browsed the piles of fruit, Lyndie picked up a peach.

The proprietor of the store, a tall, rotund, comical-looking balding man wearing a white apron that came down to below his knees, swooped in out of nowhere and quickly lifted, if not snatched, the peach from Lyndie's hands. Then he pointed to a sign above the bin of fruit that read LET LOUIE PICK YOUR FRUIT. He clearly did not want customers pawing over his produce. His face produced an unctuously insincere smile as he pronounced, "It's more sanitary to do it this way. That's why I say 'Let Louie pick your fruit.'"

The way he spoke those words—"Let Louie pick your fruit"—was unique, and to our ears endlessly funny. He enunciated each word with such precision and heartfelt devotion to the principle embodied in the phrase that it would have been impossible not to obey his command, if not fall under his ingratiating spell altogether.

Louie's phrase stuck. Whenever one of us three wanted to crack up the others, we'd simply say, "Let Louie pick your fruit." We'd say it with a characteristic emphasis, as "Let *Louie* pick your *fruit.*" Many years later I'd see the poetry in that simple sentence, the alliteration of the *l*'s and the scansion of the syllables. But it was not just the lyricism of the line that made it stick; it was the entire character sketch of Mr. Louie that the three of us instantly filled in during those few minutes in his presence. We believed we knew Mr. Louie very well, even if we actually knew next to nothing about him.

We'd get an intuition, a feel for a person or a situation in a heartbeat. Looking closer at people and situations, we drew remarkably similar conclusions. It was as if we shared a set of lenses ground exactly the same. It's why we loved being with one another, and why being out in the world *without* one another made us—or at least me—so nervous. How could I get by with no one to understand life exactly as I did?

Most people would forget about Mr. Louie and move on, but we were drawn to the Mr. Louies. We *liked* weird. Probably because we felt kind of weird ourselves—Jamie secretly dealing with being gay, Lyn knowing she was not Miss Popular in her high school, and me living with my own uncertainties—we felt an affinity for odd ducks.

Years later, when I learned that I had ADD as well as dyslexia, I'd discover that it is common for people who have those conditions to grow up feeling different, not in a good way, on the outs—if not outright misfits, then at least ragged around the edges, trying to figure out how to be smooth.

At the time, I didn't know about Jamie's and Lyn's struggles. I didn't know how brave Lyndie had to be to stay in Chatham, or how much abuse she dealt with. All I saw was Lyndie the queen, soon requesting she be called Lyn. I obeyed instantly, lest I incur her wrath.

My brother Johnny, two years older than Lyndie, would tease her now and then, calling her Bossy Jossie, as her true first name was Josselyn. She *hated* being called Bossy Jossie, and as soon as she was old enough to manipulate Johnny into submission, she put a permanent end to the use of that moniker.

Lyndie was the one who would come up with group projects. For example, at her urging, Uncle Jimmy bought a used car for us when Lyn turned sixteen and could drive. It was a beat-up old heap that couldn't have cost more than fifty dollars. Under Lyn's guidance, the three of us set about painting the car. We knew nothing about painting a car, but after a couple of days' work we had put paint on metal and changed the color of the car from dirty gray to clean, bright blue.

She always had some idea cooking. For her eleventh grade biology project at Chatham High School, she followed the progression of a chick embryo through various stages of development, carefully excising the developing chick from the yolk for display, until finally one fertilized egg was allowed to hatch on its own.

The question then arose, what to do with the newborn chick? Killing it, which Duckie advocated, was out of the question for Lyn. Uncle Jimmy, of course, favored whatever Lyn wanted to do with the chick, so a new member of the household took up residence in the house on Kettle Drum Lane. Uncle Jimmy suggested we name the chick Falstaff, with which Lyn enthusiastically agreed, having just studied *Henry IV.*

Duckie hit the roof. "How the hell do you housetrain a chicken? I am damned if I am going to go around picking up chicken shit!" Duckie rarely used profanity, so this was clear evidence she was ready to go to war.

Lyn was always ready to do battle with her mother. They had some truly nasty fights and said horrible things, especially Lyn to her mother. It didn't bother me that much, because I knew they loved each other, but the arguments really shook Jamie. He was traumatized by their fighting and, in spite of harboring a truly vindictive and unforgiving part of himself, developed a lifelong avoidance of overt conflict. When Lyn said to her mother, "Why don't you go fuck a tree?" Jamie looked at the floor as if he wanted to melt into it.

When Duckie started to walk toward Falstaff with a knife, she was stopped in her tracks by ever-resourceful Uncle Jimmy, who declared, "I have an idea!"

Duckie halted her march toward execution. Lyn paused. "How about diapers?" Uncle Jimmy proposed. "They work on humans, why not on chickens?"

"Jimmy, you can't be serious," Duckie said.

"It's a great idea," Lyn said. "Jamie and Neddy will help me put the diapers on and change them, won't you?" We both nodded dutifully, even though I barely knew what a diaper was, much less how to put one on a chicken or take it off.

So it came to be that a chicken wore diapers in the house on Kettle Drum Lane. Fifty years later, the son of the Penningtons, the people who lived across the street, bumped into me at a party. "Are you related to the Hallowells who had a chicken who wore diapers?" he asked.

Knowing there could be only one, I nodded.

9.

I started kindergarten in the public school in Chatham. After Dad moved out of the house and up to Wianno with Gammy, we remained living on Seaview Street, near Uncle Jimmy, Duckie, Jamie, and Lyndie, just across the golf course and down the hill from the McMullens' house, the people who owned the Chatham Bars Inn, and whose twins, Helen and Bess, I went to school with. Next door to us was my friend, Terry Pickard, whose father had a pile of lobster pots on the grass between our two houses. We'd turn those lobster pots into cages for our made-up games.

I was a curious, imaginative kid. The grown-ups called me the Question Box because I would wear them out with endless questions until they finally had had enough and said, "Stop! I simply can't answer any more of your questions." They didn't say it in a harsh way, just a weary way. So I'd go away for a while, always to come back with many more questions.

I did not have a typical brain. My mind went off on tangents all the time, and in addition to flooding my family with questions, I also found so many things funny that I would laugh and giggle excessively. It was a habit I had to work to control.

It was in first grade that I initially encountered my problems with reading. I didn't learn to read as quickly as other kids. Back then if you were slow to read there were no learning specialists to be referred to. Your "diagnosis" was obvious: stupid. And the "treatment plan" was equally clear: Try harder.

In some schools those in charge used shame and humiliation to motivate slow readers: standing in the corner, or wearing a dunce cap. In some schools, kids would get spanked. If none of that worked, the diagnosis was that you were *really* stupid. The IQ tests and what formal assessments there were actually used terms like "moron," "idiot," and "imbecile" to identify differing levels of stupidity or retardation. But that was OK. People said there was always a place in the world for the really stupid.

I was lucky. I was never given any of those tests. I also had a first grade teacher by the name of Mrs. Eldredge, who knew that there was more to little boys and girls who were slow readers than being stupid, and there were better ways to help them than by shaming or punishing them. She had no special training in reading that I know of, but she'd been teaching first grade for a long time.

Mrs. Eldredge was quite plump and very kind, but she could also be tough, so no one messed with her. During reading period, with us sitting at round tables reading aloud from the "See Spot Run" books, when my turn would come and I would have trouble decoding the words, Mrs. Eldredge would take a seat beside me and put her arm around me. She didn't excuse me from reading, but her arm took away any embarrassment I might have felt as I stammered and stumbled over the words. None of the other kids laughed at me, because I had the enforcer sitting next to me.

Her kind nature and soft arm took out of the process of learning to read the damaging disabilities: fear, shame, and believing something was wrong with me. Mrs. Eldredge's arm is what we would now call my IEP, my individualized educational plan, and it was the best IEP ever devised.

Had I had a different first grade teacher in a different school system, I likely would have acquired the toxic disabilities of shame, fear, and selling myself short. I will always be grateful to Mrs. Eldredge and the Chatham public school system for giving me such an excellent start.

My father would tell me later that year that I was what's called a "mirror reader." When I asked him what that was, he spoke a long word that I didn't understand. It was probably "strephosymbolia," a term commonly used back then for people who made reversals and read some words backward. Now it is all subsumed under the heading of dyslexia, or a reading disorder, which I do have.

When my dad told me what was wrong with me, and I didn't understand the long word he used to explain it, I asked him again what it meant.

"It means you read some words as if they were in a mirror. You read them backwards."

I asked him why I did that.

"You were just born that way. But don't worry, you're learning to read all the same."

"But how can I read them if they're like in a mirror?"

"It's just how your brain works. But like I said, don't worry, you will still read just fine."

For a while after that I'd hold pages up in front of mirrors to see if I could read them better that way, but I couldn't. Once I realized that a mirror didn't help, I pretty much forgot about it and just tried my best to read as well as I could.

Throughout my schooling I knew I was a slow reader, but that didn't hold me back. I learned to read well enough to excel academically and major in English in college, graduating with high honors. That would not have happened without Mrs. Eldredge.

In addition to reading slowly, I also went about solving problems differently than other people, often inventing my method on the spot. I preferred free essays rather than being assigned a topic. I liked to think of off-the-wall examples rather than common ones. So, if asked "Name something you throw," instead of saying "A ball," I'd likely as not say "A plate of spaghetti." I wasn't trying to be a wise guy, I just liked being playful, going off in unexpected directions. I knew I was different but it didn't bother me, in part because eccentricity was such an accepted part of my extended family.

I also had a really rich fantasy life. At age six I'd imagine I was the boss of the world, and I'd be ordering armies around as I walked the streets of Chatham. Even though I felt safe in Chatham, I would imagine there were bad guys ready to pounce, hidden in various houses around town. I was really afraid of the dark and usually pulled the sheet up over my head, even though Chatham was as safe a town in the 1950s as a town could be. We never even locked the doors on our cars or houses.

When I was in college, I asked Uncle Jimmy why he had decided to go into farming. He was sipping his usual mug of tea, sneakered feet up on the kitchen table, tilting back in his wicker-seated ladder-back chair, smoking the pipe he always seemed to have in his mouth.

He loved these kinds of questions, because it gave him a chance to philosophize. He loved to talk about the meaning of life and why people do what they do. "I tried the banking world for a while, you know that, don't you? I hated it. It wasn't me. There were lots of people trying to be someone other than who they really were, and I just did not want to do that. To thine own self be true."

"Yeah, so I've heard," I said.

"Farming seemed real to me. You reap what you sow. There's no faking it."

"And when you had to sell the farm, why did you pick Chatham and the bowling alley?"

"Well, the girls, your mother and Duckie, grew up in Chatham on Minister's Point. So they loved the idea of Chatham. And your dad and I grew up in Wianno, when we weren't in Chestnut Hill. The bowling alley was a lark. It was a hot business in the fifties, and I've always liked machinery."

He didn't add that in addition to alcohol, back then he'd used heavy doses of the various antianxiety agents of the day, like Librium, to get through. He never talked to me about that, but we all knew it.

At the end of that conversation I told him how much I admired him for staying true to himself, and that I would try my best to do likewise.

The older I got, the more I sensed I was different in ways like Uncle Jimmy, but I had no terms or diagnoses to attach to it. The reading issues became no big deal, just an inconvenience, if that, because I always got top grades.

But the feelings of insecurity stayed with me along with pockets of depression we now call low self-esteem. Although invisible, they set me apart. Like Uncle Jimmy, I did my best to hide all that from the world, as well as a growing awareness that my mind really did not function the way most kids' did. Not knowing how to explain it, I didn't try. But the enduring gift I got from my family is that being different did not cause me to feel ashamed.

10.

Maybe a year after Dad left us, I found myself playing in the backyard of the man I would come to call Uncle Unger. He had a beat-up, slightly deflated leather soccer ball—I'd never seen any kind of soccer ball before—and Unger Stiles was playing keep-away with me in his spacious yard. I ran after him, happy as could be, trying to kick the ball from his control. He wore fraying gray flannel trousers, a worn-out white Brooks Brothers button-down shirt, and dirty gray Keds. He looked pretty old to me but he was quick on his feet.

I liked him a lot. I had no idea who he was or why I was playing with him, nor did I ask or care. He was fun, and he paid attention to me, that's all that mattered. He was just a man Mom had brought me to visit. No mystery, no big deal, as far as I was concerned. Just fun.

I was wearing my usual shorts and striped jersey. This new man had leathery skin and graying hair combed straight back, a look I'd never before seen, as I was used to hair parted on the left or right in men, like my dad, or crew cuts in kids and Uncle Jimmy. Unger Stiles moved around the yard easily, crouching and dodging, swaying side to side to trick me as I tried to kick the ball away from him. Every now and then I'd succeed, and the ball would shoot off into a briar patch of wild blackberries, purple thistles, serpentine crawlers, and convoluted creepers all studded with the nastiest of tiny thorns that did not protrude far enough to give you visual warning but scratched you nonetheless upon every visit into their midst. But I gladly plunged into the tangled mess and fished out the ball, not at all minding the scratches, to bring the ball back and keep the game going. Playing with Unger was great fun. I found myself liking him more and more. That Dad had left didn't seem to matter so much.

After a half hour or so of "soccer," Mom, "Uncle Unger," and I went inside, where he asked Josephine, his colored maid ("colored" was the word used in Chatham back then), to give me a Coke while he whipped up some cocktails for Mom and himself.

Soon he was sitting in his easy chair by the fireplace and entertaining me by holding a cigarette between his upper lip and his nose. Doing this made his face look awfully funny, and I giggled. "You try it," he said.

Within a few minutes I was able to do it as well. I couldn't wait to show Jamie.

Soon, a visit to Uncle Unger's became such a treat that my mother would use it as an inducement to persuade me to go to doctor or dentist appointments. Even though Jamie would at the same time try to scare me into believing those appointments would include intense pain, shots from the doctor, and some diabolical procedure that involved a jackknife, the thought of visiting Uncle Unger was enough to get me to go anyway.

Before I turned six I was sitting in a grand living room in the mansion of Junior Howes, whom I called Uncle Junie, watching my mother walk down a makeshift aisle to marry Uncle Unger. Junior, one of the wealthiest men in the area, and his wife, Dot, who was always really kind to me, were two of Uncle Unger's best friends.

Jamie, who attended the wedding, was about ten years old and remembered it well. He would later tell me he thought it was really stylish and fun, creating a look like that of the movie *High Society*, starring Bing Crosby, Grace Kelly, and Frank Sinatra, which came out the same year Unger and Mom got married.

Being so young, just a few months short of my sixth birthday, I didn't really grasp the significance of the wedding or even vaguely understand how dramatically this event would change my life for the worse. Instead, I was happy as a Chatham clam. I do remember marveling at the opulence of the Howes estate. I was mostly concerned with finding Jamie and eating the hors d'oeuvres. No one could have prepared me for what was to come, because no one knew.

After the wedding, when I wasn't living on Kettle Drum Lane, I lived in the house on Bridge Street. When I was living with Jamie I'd walk to school with him, but when I was on Bridge Street I took a yellow bus.

I didn't have any friends in the Bridge Street neighborhood, so I would occupy myself by reading in my room, hanging out with Uncle Unger's maid, Joey, and her friends, or walking around in the fields and woods out back. I'd take a bow and arrow and try to shoot a rabbit, but I never bagged one.

Every now and then Jamie would come to Bridge Street to play, but usually I would get driven down to Kettle Drum Lane, or if it was still light out I'd ride my bike. Sometimes I'd stay at the Kettle Drum Lane house for days or weeks.

Lyndie and Jamie started to resist coming to see me on Bridge Street because after Uncle Unger married my mother, my friend, the man I had so liked, changed. There were many more martinis and fewer soccer games in the backyard, if any. A mean streak flared up. One incident stands out, a harbinger of many others.

Uncle Unger had gone to Harvard but had dropped out before graduating due to some dispute with the administration. He did, however, boast about attending Harvard, and he treasured the highball glasses he owned with the red VERITAS crest embossed on them. He regarded these glasses as priceless, even though I would later learn they could be purchased at the Harvard Coop for small money. Obviously their value to Uncle Unger lay far beyond their cash value.

One day Jamie accidentally dropped and broke one of those prized glasses. Uncle Unger launched into a tirade, yelling at Jamie, cussing at him, and raising his hand as if to hit him. Jamie must have been ten at the time. I remember his starting to cry, and my feeling that I should do something, but I had no clue how to quiet this raging man whom I'd recently so adored.

It was about five o'clock in the afternoon, the rays of the sun slanting in through the windows of the house, highlighting the hair in Uncle Unger's ears and nose. The people in my family could be strange, but they were never cruel, and they never struck fear in me.

But in that moment, the sun accenting hair where I didn't know it could ever grow, I felt what I'd never felt before (but would often feel again): I was terrified by a person I looked up to.

I should have known then, even so young, what I would come to know as well as I knew my name: the signs of Unger's having dipped into gin. Gin drew out the rage. I remember the fear, my concern for Jamie, confusion over how Unger could suddenly become so mean, and my wishing for life to be different. I wanted Dad to come back, or for me to move permanently to the house on Kettle Drum Lane.

Unger retreated back to his chair by the fireplace and Jamie and I fled to the basement to play. We both felt really nervous. I remember Jamie trying to console me instead of vice versa. Even after being terrorized by Unger, Jamie tried to look out for me.

After he went home I told Mom what had happened and she said that Uncle Unger really loved his Harvard glasses, but that he should not have been so harsh with Jamie, especially since dropping the glass was an accident.

Later that night, eavesdropping from the top of the stairs, I heard Unger and Mom fighting about what had happened.

"He's a brat," Unger spewed. "I would have spanked him if he'd been mine. He's not welcome here anymore."

"That's not fair," my mother protested. "He's Neddy's best friend."

"Well, Neddy needs to make some new friends. Your sister Miffie and her whole family, Jimmy and Lyndie and Jamie, they are all inbred and twisted. It's not normal. That's what happens when two brothers marry two sisters. Mark my words, Neddy will not turn out well. Look at his brother Johnny. All he wants to do is go down to Kettle Drum Lane and fuck Lyndie."

"Unger, be quiet," Mom said. "People can hear you. What you're saying isn't true and it's way, way out of line."

"Don't tell me what's out of line. You're totally naïve. Don't you know what the score is, *sweetheart*? Now with Ben out of the way, Jimmy has his eye on you, for you to get in bed with him and Miffie. Why else would brothers marry sisters? It's all in the family. A true family affair."

"That's disgusting," Mom said. "And wrong."

"So say you. Don't you see how Jimmy looks at you? He says he loves you like a sister and he'll look out for you. *Hah!* Some kind of sister. I guess we'll see some day what kind of 'looking out for' that actually turns out to be."

"How can you talk that way? How many of those martinis have you had?"

"Enough to tell the truth. Can't you see what's right in front of your eyes? You're so innocent. Your whole family are a bunch of innocents. It's truly amazing. And what do you think is up with your father and Miffie? What do you think all those trips to New York and the Rainbow Room and Guy Lombardo are all about?"

I had no idea what he was talking about. Duckie and Skipper loved to dance. Duckie would try to teach me how to waltz from a very early age.

"You're drunk," Mom said. "I can't listen to this."

"Fine, fine. Have it your way. Just shove off and leave me alone."

"Shove off" was one of his favorite conversation stoppers. Sometimes he elongated it to "Shove off, coxswain, and make your regular trip and return," punctuating it with a flip of his wrist. It was a line he took from the Navy: captains of ships would say it to instruct what were called the coxswains of the dinghies that would come up alongside to transport people ashore. I would hear that phrase hundreds and hundreds of times again. I can even hear it today, with the drunken slur in which he usually spoke it.

From my perch on the stairs, listening to words I didn't understand but knew were bad, I felt confused. How could this man whom I had liked so much become so mean?

Their words were both riveting and terrifying. Why did I listen to Uncle Unger and Mom fight that night, and why did I continue to listen so many, many nights after that? Why didn't I plug my ears or just go up to bed and put the pillow over my head?

I guess for the same reasons people stand and watch their houses burn down.

11.

When my mother and Uncle Unger were together in Chatham, Joey often looked after me. She was fun to be with, she taught me how to cook, and she became my friend. I often didn't know where my mother or Unger was but didn't care because I had Joey. She was nice and really pretty and always very peppy.

One day we stopped on Shore Road, and Joey told me to wait in the car while she ran across the street to pick something up. Being curious if not downright nosy, I watched her intently. I loved Joey and wanted to track her every move. In a moment, I saw her friend Topper emerge from his little gardener's shack, meeting her at an opening in the tall green hedge. Joey often stopped to see Topper, just for brief visits.

Topper looked odd to my young eyes, but in a good way. He was brown like Joey, but he looked much, much older. He was fit enough to garden but still appeared frail, as if he might just fall to the ground if Joey patted him on the back. I'd never seen anyone with his look, almost as if he had no blood in him. He had jet-black hair, thick with cream or oil that matted it down like a helmet and made it shine. When he saw Joey—or anyone, judging from the many different times I'd see him—he always broke into a wide smile, which, a year or two later, when I learned who Louis Armstrong was, I would see as an exact replica of his.

Joey and Topper exchanged a few words, as well as envelopes, before Joey gave Topper a kiss on his cheek and ran back across the street in her blue and gray maid's uniform to the classic Ford woodie station wagon my stepfather had her drive.

She got in the car and said, "Only one more stop, then we go home and do some cookin', hon." She seemed happier than usual. I smiled, licked my lips, and said that was great, as I loved to cook with Joey.

I didn't know it at the time, but I would later learn that Joey was heavily into hard drugs, like smack, cocaine, and meth, as was Topper, who was her dealer. He was more than her dealer, though. They were fast friends. And even though Joey may have been high most of the time, she was always attentive to my needs and safety and couldn't have been kinder to me if she were of my own flesh and blood.

Unaware that I had just watched one of the many drug deals I would unwittingly witness during my childhood years in Chatham, I sat back and listened to Elvis sing "Love Me Tender" on the car radio as we wound our way into town to pick up some chicken at Bearse's, the grocery store my older brother Ben used to work at summers before he went off to Lawrence Academy and then Annapolis. Turns out Roy Bearse, the owner, paid Ben a little extra to cover for him when he'd leave early to drive to the dog track. There were many such little secrets in the town.

Provisions soon in hand, we drove back to the stately white house on Bridge Street. My brother John hated Unger and had gone to live with Duckie and Uncle Jimmy, and I can't remember where Ben lived at the time.

I sat on a white stool in the kitchen while Joey started the fried chicken. I watched intently as she dipped the naked chicken pieces in buttermilk, then in a flour mixture, her secret recipe, and then laid them out on a platter next to the frying pan as the Crisco heated up. "You see, hon, Mr. S. asked for cold fried chicken for dinner tonight, but the chicken has to get hot before it gets cold, don't it now?" She gave me a little wink, then disappeared up to her room for a moment.

I assumed she had to go to the bathroom. "If the grease starts to spit, pay it no mind, I'll be right back. Get yourself a Coke."

You bet. I popped open a bottle of Coke I took from the fridge and took a long drink of it before Joey reappeared, practically dancing into the kitchen, unbeknownst to me high on the drugs she'd just bought. I just thought she was happy.

"After I cook this up, I'm gonna take you out to see Vera and Pokey, because Vera wants me to bring her some of the chicken. Don't be telling Mr. S. that I share food with her, OK?"

"Can I tell Mom?" I asked.

"Best not to, OK hon?"

"Sure," I said, not understanding why it would matter. I was an uncomprehending witness to all sorts of doings during my early years, which only adds to the spell they cast for me today.

A gravel driveway led from the street down a slight hill and around behind the house into a garage underneath. Most mornings Uncle Unger would go down to the garage and rev the daylights out of that woodie. When I heard it for the first time, I thought the car was going to explode or the garage catch fire, but Uncle Unger later explained he was just "blowing out the carburetor." The sound was similar to what I'd hear him doing to his own lungs every morning, coughing until I imagined him bringing up his insides. I'd never heard anything like it.

"It's because he smokes so much," Joey told me one morning while making my favorite lemon meringue pie. "He has to bring all the junk up the next morning."

"Is there blood in it?" I asked.

"Well, now, aren't you just the little curious one? You oughta be a doctor. Honestly, child, you are *the* most curious boy I have ever met."

"Why does he smoke if it makes him so sick in the morning?" I asked, while also eyeing the lemon filling being patiently prepared in a double boiler, letting the egg yolks thicken first. Once it was ready—which always seemed like an eternity—I would get to lick the pan after Joey poured the filling into the crust.

"It's a bad habit, smoking. I smoke, too, you know. But not nearly as much as Mr. S. When he starts drinking, then he chain-smokes. That means one cigarette after another. I'm sure he doesn't like all he has to go through the next morning, though."

"I'm never going to smoke," I said.

"I hope you're right," Joey said, as she offered me the bowl to lick. I was wrong. I smoked senior year of high school and didn't quit until Lucy, our first child, at age three, said to me, "Daddy, when are you going to stop sucking on cigarettes?"

Back of the house was a carefully kept emerald-green yard. I'd never seen greener grass until the magical moment I first walked up out of the bowels of Fenway Park and beheld grass as green as grass could be.

Unger's backyard was bordered by well-tended lilac bushes and swaths of lilies of the valley, which gave way to spontaneous black-berry patches, and then a tangled panoply of various wild, knotted growth, common weeds, but lush with the clashing colors Nature awards her uncultivated, unplanned vegetation, like purple strife, bittersweet, and jimsonweed. And beyond all that lay the currents and salt airs of Monomoy Sound.

When I was feeling unhappy, Uncle Unger would teasingly try to cheer me up by suggesting I go out back under the lilacs and dig worms. I didn't get the joke, nor did I ever take him up on his suggestion.

The scent of lilacs and lilies of the valley, as well as the idea of digging worms, still bring me back to that yard, as well as to the memory of Mom sitting on the grassy slope behind the driveway, sewing nametags, with a needle and thread, onto clothing I was taking to summer camp.

I also remember her and Unger, and sometimes his two grown-up daughters, Allison and Hope, telling jokes and laughing as they drank bourbon on the rocks with a splash. There would be a silver ice bucket and water pitcher perched nearby in the grass while they snacked on hard-boiled eggs with a dash of salt—saltshaker in the grass as well—and Ritz crackers spread with Underwood deviled ham, its paper wrapper with the little red devil holding a pitchfork still on the can.

I never understood the Stileses' jokes, but it was fun that they laughed a lot. I also didn't know then about the tragedy that had befallen them before my mother met and married Unger. Even as they hid it, I don't believe that the three of them—Unger, Allison, and Hope—ever fully recovered from it. They lived fully, drank a lot, ate the great food Joey cooked, played music, danced inside and outside, and seemed for all the world to be happy as could be.

The girls would go on to excel in college, and Hope would marry and have several children. I lost track of Allison, but my brother Ben tells me she never did marry.

And those nametags. What a labor of love, sewing nametags one at a time onto countless socks, underpants, jerseys, shorts, and everything else I had to take with me and might lose at camp.

Six years old was young to go to sleepaway camp for two months in New Hampshire, a three-hour drive from Chatham, to Lake Winnipesaukee, but my brother Johnny would be at Camp Kabeyun as a counselor, and Jamie would be there as well. Johnny would ignore me, but as long as Jamie was there I'd be OK.

I ended up loving Kabeyun, going there for five eight-week summer sessions, learning to shoot a rifle and a bow and arrow, paddle, steer, and portage a canoe, climb mountains, build a fire with no matches, wash tin plates in cold streams, and pitch a tent. During one of those summers a man named Fidel Castro overthrew a dictator named Fulgencio Batista, about which I knew nothing except that the head of Kabeyun, John Porter, a spunky seventy-year-old lefty and his chain-smoking, bridge-playing, highly intellectual wife, Anna, along with most of the counselors, let out loud cheers when the news of Castro's victory reached us on Lake Winnipesaukee.

12.

I sat in the passenger seat as Joey drove. We were heading to Harwich to visit Vera, Joey's cousin, and Vera's daughter, Pokey, as well as their friend Bernice, and whichever men were hanging out at Vera's place.

She always had good snacks to offer, and the yard was full of chickens, which I had fun chasing. I also liked Pokey. She was much taller than me but about my age.

When we got there, Vera and Bernice were in the kitchen, smoking as usual. No men today, and no other children except Pokey. After Joey gave Bernice a little envelope, that looked a lot like the one she'd taken from Topper, she lit up a cigarette, too, as Vera called, "Pokey, time for your bath."

There was a big aluminum tub in the kitchen, which Vera had filled with hot water. "Neddy, you ought to take a bath with Pokey."

I stared at the tub in disbelief.

The ladies laughed a little at this idea, but I could also sense they meant it. "C'mon, jump in the tub."

Pokey came in and, without so much as a by your leave, stripped off her clothes and climbed into the tub. *Wow.* I was excited because I had never seen a girl naked before. I didn't see much beyond a fleeting glance when she got into the tub, but I was eager for more. Still, I felt beyond shy about taking off my clothes. No way was I doing that.

"It's OK," Bernice said to the others, "they're still babies. I don't think we need to worry about any hanky-panky."

Vera giggled at the thought, and Joey chimed in, "C'mon, Neddy, take your bath now and you won't have to at home."

"I don't want to."

"C'mon," Pokey said. "It'll be fun."

I looked up, I looked down. I didn't know what to do.

"C'mon, Neddy," Joey said. "You be gettin' naked with girls for the rest of your life!" They all laughed.

I don't know what gave me the courage, I guess Joey's urging was all I needed, because in a moment I was naked in the tub with Pokey. The women took their cigarettes and iced tea—or whatever was in those tall glasses—into the sitting room, leaving Pokey and me alone in the tub.

Even though she wasn't older than me, Pokey seemed more mature, less silly. She could tell I was nervous, so she said, "I bet you ain't never seen a girl naked before, have you?" And then she laughed, not like laughing at me, but like she was saying it was really OK not to have seen a girl naked.

"I guess not," I said. "Well, I did bring my stepsister a drink when she was in the bathtub, but she was covered with soapsuds, so I couldn't see anything."

"You want to see me?" Without waiting for an answer, Pokey stood up so I could see what was, or wasn't, between her legs. I looked up at her face, then down at her vagina, staring at it for what seemed like forever, except it wasn't forever.

Before I knew it, Pokey giggled and sat down again in the tub. "Now it's your turn," she said and giggled again.

As if it were simply what I had to do, I stood up and let Pokey see me, then quickly sat down, making a splash.

"What's going on in there?" Vera called.

"We're just washin'," Pokey called back, giggling again. Pretty soon we were splashing each other and making a racket, so Vera called in, "You two better quiet down in there or I will come in and give you what-for!"

"Shhh!" Pokey said to me.

"What's what-for?" I asked.

"It's a spankin'," Pokey said, "and you surely do not want one from my mom!"

"OK," I said, still giddy over what I'd just seen, "I'll try to be quiet."

When we finished our baths, dried off, and got dressed, Vera gave us each a drumstick and a scoop of potato salad.

Driving home, Joey asked, "Why you so quiet?"

"No reason," I said. But of course there was a huge reason. I didn't know what that space with what looked like vertical lips actually was on Pokey, but I guessed I'd find out someday. It meant Lyndie had one like it, only not brown, and so did my stepsisters, and all the girls in my class.

I also knew I liked seeing it. I hoped Pokey and I could take baths together often.

I turned over lots of questions on the car ride home, but I didn't know who to ask them of. *Jamie*, I thought. *Jamie will know.*

That turned out to be the only bath Pokey and I ever took together. It was one of the sweetest, most educational experiences of my life.

13.

Josephine, Sammy Sanders is coming over this afternoon. Let's do the caviar today." Hearing Uncle Unger's words from downstairs, I looked up from what I was reading while lying on my bed. It was a short book about Abraham Lincoln, assigned by Mrs. Brown, my second grade teacher, whom I adored. I may have been a slow reader but I still loved to read. However, the prospect of Uncle Sammy joining Uncle Unger for backgammon was enough to pull me away from Honest Abe.

Sammy Sanders owned a hugely successful seafood restaurant right on the water a few towns up Route 6 from Chatham. He was nice to me, he was funny, he was the absolute fattest man I had ever seen in my life, and he wore loud Bermuda shorts—yellow, green, or pink—along with loafers and no socks.

Coming downstairs when I heard Sammy arrive, I saw Joey greet him, then saw Unger come out from the living room, saying, "Welcome to the land of opportunity!"

"Fuck your opportunity," Sammy said. "With the money you've taken off me, I could open three new restaurants." People swore a lot in the Bridge Street house, which, looking back, is odd, because Uncle Unger always stressed that the most important thing in life was to be a gentleman.

"Being a gentleman means you are always considerate of others," he'd say to me. If I'd questioned him about the swearing, he probably would have said that it puts other people at ease, which is why a gentleman can swear.

"Josephine is preparing a plate of caviar," Unger said, ushering Sammy into the living room. I followed.

Sammy took his customary seat in a wing chair next to the fireplace. The backgammon table was already in place between Sammy's chair and Unger's.

Before Unger sat down, he said, "I think we need a fire." He'd already laid it, so all he had to do was light the pumice head of the Cape Cod lighter that sat in a brass canister on the hearth. Soon a blaze would warm us.

I sat on the oriental rug on the floor, legs crossed, and watched. For some reason Sammy liked having me around, and he didn't mind my questions. Unger called me the Question Box, but he put up with me. Also, I'd run into the kitchen and serve as errand boy as needed.

Walking over to the mahogany table that served as a bar, Unger said, "Are you ready for some silver bullets?" Early on, I learned that meant martinis.

"Of course!" Sammy replied.

Glasses in hand, Unger went back to his chair, sat down, and began a ritual I would witness many times in the seven years he was married to my mother. Like a chef preparing a signature dish, he'd carefully chill each glass by holding its stem in his thumb and forefinger so he could agitate the glass to make the chip of ice he'd put in it spin around inside, rising up the side of the glass to the rim, chilling the entire glass until the ice chip melted. Then, as if casting a fly-fishing rod, he'd flip the resulting water out of the glass, creating a long and graceful liquid arc, onto the rug. Once he'd done this twice, he had two frosty glasses ready to be filled, so he asked me to go get the silver shaker in which he'd already mixed the gin and vermouth, and to bring along two twists of lemon peel as well, which he'd precut and left sitting in a small Canton china dish on the bar. Jumping up, happy to join in, I did my job.

Taking a generous sip of his martini, Sammy let out an "Ahhh," purring in sheer delight. Years later, I would learn the singular pleasure a drinker enjoys when he takes his first sip of that day's first martini. "Today's my lucky day, Stiles. Just you wait."

In a few minutes Joey called out to me from the kitchen. I went in, and she gave me a platter with caviar in a small crystal dish sitting in a larger crystal bowl of crushed ice with a mother-of-pearl spoon for the caviar lying atop the ice. This larger bowl was surrounded by lemon wedges and toast points, as well as small piles of chopped egg white, chopped egg yolk, capers, and chopped red onion. "Can you bring this in to Mr. S. and ask him if he need anything else?"

Like a little ringbearer, I carefully carried out the platter and put it on the mahogany side table next to Uncle Unger's chair. All but drooling, Sammy slapped his knee and purred again.

In moments, they were eating, drinking, and shaking dice. Mom was off visiting Duckie, as she disliked these afternoons of drinking and backgammon. I watched, knowing what would happen but always finding it funny. I'd wonder why Sammy was so stupid. But still, he was nice and he was funny, so I liked him a lot. As usual, Uncle Unger let Sammy win the early games while he gloried in getting drunk. Uncle Unger drank just as much but never got nearly as drunk.

It was the key to his success. Once Sammy was good and soused, Uncle Unger would start doubling, which Sammy couldn't resist accepting and doubling back. I watched until I'd seen enough to know that once again Uncle Unger would win big. I went back to reading more about Abe.

After a few hours the doorbell rang, on cue. Joey answered it and said, "Come in, Gerald." Gerald was Sammy's aide, the man whose task it was this day to assist Sammy to the car and drive him home. Sammy was so drunk he couldn't walk without both Unger and Gerald propping him up. But he was a happy drunk. "You're a lousy, stinking, cheating bastard, Unger Stiles, but I love you all the same."

"And you're a gentleman and a scholar, Sammy Sanders," Unger said as he gave him a kiss on his cheek.

14.

A year or so after she married my stepfather, my mother asked me if I would mind if the three of us moved from Chatham to Charleston, South Carolina.

All of seven years old, I stepped up and said yes, I would most definitely mind, that I actually thought it was the worst idea I'd ever heard of. Lyndie, Jamie, Duckie, Uncle Jimmy, and Joey, they were my family, people I absolutely did not want to leave.

A few weeks later, bags packed, we were being driven to Boston to catch a train to Charleston.

Leaving Cloverluck Farm had been no big deal for me, but moving to Charleston sure as hell was. I should have learned then and there that you can't depend on anything. But the paradox is, I kept depending and trusting just the same. Some people trust no one; I trusted and still trust everyone. I saw what a mess life can be but somehow didn't see it as a mess, just as life in progress. Things were never so bad that I didn't always have some solid person nearby who kept an eye out for me.

My protectors provided moorings in the surge, points of stability that to this day I count on for balance and support as much as the bones in my feet. I never knew when or where I'd find these people, but miraculously they always appeared.

The move to Charleston began innocently enough. After the car ride with Uncle Unger and my mother to Boston's Back Bay Station, chauffeured by Chatham's only taxi/limo service in what I remember was a chubby 1955 Cadillac sedan, we boarded the first sleeping car I'd ever been in. It was called a Pullman car. My mother told me that my uncle Dick, my father's sister Nancy's husband, worked for Pullman, so I felt proud to be in one of my uncle's cars, as if he owned it.

On the train, we had what was called a bedroom. It was a sleeping compartment with three bunks. I thought it was just so cool how they could fold down beds and turn what had looked like a sitting room into a room full of neatly made bunks. The *clickety-clack, clickety-clack* of the train making its way down the Eastern Seaboard while I fell asleep cut a soundtrack in my brain that's still there.

Just as I fell in love with words playing ghost, on that train trip to Charleston I fell in love with trains, and with long trips in general. Even though leaving Chatham was the last thing in the world I wanted to do, that train ride cured my sadness. I didn't know where I would be going to school, but I knew I would not be seeing Jamie and Lyndie for a long time. I had no idea what kind of house we would live in or what the people were like in Charleston or why we were moving. In spite of my saying it was a terrible idea, the rocking movement of the train and the rhythmic *clickety-clack* made me forget all those concerns and instead focus on the adventure of being in this fantastic world created by the Pullman car. That train barreled its way past my misgivings and into the welcoming darkness. The darkness didn't frighten me in the least. As far as I was concerned, it set the stage for an adventure.

After a few hours we stopped at Penn Station in New York City. I was in my pajamas, in my bunk, snug under a white sheet and blue blanket, with the lights off, looking out the window, as we slowly pulled into the station, at a host of people dressed in all kinds of clothes, walking with leather suitcases, hatboxes, and other bundles on the platform. I wondered who they were. So many people. I'd never seen a train station that was underground. But here we were, in a huge electrically-lit cave. The yellowish lights of Penn Station filled my otherwise dark sleeping compartment while I watched the many grown-ups, and a few children, bustle along.

Slowly making its stately way past the hurried people, the great and powerful locomotive gradually brought the train to a full stop. As if exhaling from its long haul, the train let out a long hiss while a throng of people filed out of the train and onto the platform as eager others boarded. Uncle Unger and my mother were in the lounge car having drinks, so I was gloriously alone. It was thrilling to watch and listen to all the action as I reveled in the panoramic sweep of the Penn Station platform. Propped up on my left elbow, lying on my side in my bunk, I felt I was watching a real-life movie. Just from watching I could feel a definite excitement in all the hubbub, the coming and going, as if these people, like me, were heading on to some new adventure in their lives.

In a few minutes, fully rested, the mighty locomotive started up again. I was back sleeping to the *clickety-clack* before Unger and my mother returned to the room.

The next morning, the three of us got up and dressed as the train sped down the eastern shore. When I peed, I asked my mother where the pee went when you flushed the toilet. "It goes out onto the track," Uncle Unger replied. "That's why you're never supposed to go to the bathroom when the train is stopped in the station. There are signs everywhere that say that." I had trouble imagining how a train full of people could pee and poop onto the railroad track all the way to Charleston, but I didn't question Uncle Unger.

We went and ate breakfast in the dining car. I was amazed at how heavy the knives and forks were, and how thick were the china plates, cups, and saucers, all white with a thin sky blue line around them for decoration. I had fried eggs, bacon, and, for the first time, grits. "You'll be having lots of grits in Charleston," Uncle Unger said. "It's a staple of the South."

"What's a staple?" I asked.

"It's something you have every day. But a staple is also something you punch into papers to hold them together. It's a word with two meanings. There are lots of those."

Uncle Unger loved words, too. He had worked for the *Saturday Evening Post* in New York City before he retired. He was retired when I met him. Jamie told me being retired meant he was rich enough so he didn't have to work.

15.

People I would come to know well met us at the station in Charleston. They were cousins of Unger's: J. J. and Chance Ravenel. One of their sons, Trippy, was about my age and became one of the few friends I made during my three years in that city.

"Hey," Trippy said that day as he ambled my way. I'd learn that "Hey" was Charlestonese for "Hi." "Wanna put a penny on the track?" Trippy asked with great enthusiasm.

"Sure. What does that do?"

"When the train leaves the station and rolls over the penny, it flattens it out like you can't believe."

Trippy took a penny out of his pocket, walked over to the track, bent down under the train, and put the penny on the rail in front of one of the smooth steel wheels. I watched carefully. There was a stark contrast between the black, greasy undercarriage of the train hovering above the blackened, sooty stones of the railroad bed, and the still, silvery rail itself that practically glistened in the daylight of Charleston, as well as the even more silvery surface of the part of the wheel that rolled on the rail.

Sure enough, after the conductor, holding on to a silver bar on the side of the car, with one leg on the train and one leg out, cried in a long, loud drawl, "Booooaaaard! Awwwaaall aaaabooooard!" and the train, which I had so relished riding on, hissed and gathered up its haunches, gradually pulling out of the station and leaving us standing there on the concrete platform with our bags and our welcoming troupe, what was left of Trippy's penny was no longer a penny at all. It looked as if it had melted right there on the rail, like a butterscotch sundae left out in the sun. As I was about to be, it had been squished by enormous force into a totally different shape.

"Here," Trippy said, giving me the transformed penny, "this is for you. Welcome to Charleston."

I had no idea what was in store for me, but I figured I'd find a way to live without Lyndie and Jamie in this new city. We rented a beautiful house at 75 King Street, with high ceilings and fancy furniture.

Most mornings I'd wake up to a loud cry it took me a while to understand. "Shrimpy-raw-raw-raw! Shrimpy-raw-raw-raw!" Once I knew what the woman was calling out, it was easy to understand it, but until I asked our maid what on earth that woman was yelling and why, I had no idea. "Go out on the street and look," Victoria, one of our two maids, said.

Sure enough, outside was a woman, who looked a lot like Victoria, pushing a green wooden cart filled to overflowing with shrimp. Maids, mothers, and whoever else wanted to hustled out of houses and bought the shrimp fresh off the boats. The way Victoria cooked them, all crispy with a bread crumb coating, was to die for. I've never had them done like that since. They were the best shrimp I ever ate, or, as one of Victoria's friends said, "best shrimp you ever et."

As tasty as those shrimp were, the stink of Charleston's pluff mud was equally foul. I can still smell that pluff mud. Found in the salt marshes in the low country, it is a deep, gray, gooey mud created by the decomposition of the cordgrasses along with the rich marine life the tides bring in, crabs, shrimp, and various fishes, all worked on by that master creator of stink, anaerobic bacteria. When the wind blows from a certain direction, parts of Charleston reek of the hydrogen sulfide those anaerobes put out. You get used to it, even like it, the way some people like the smell of a stable, but it stinks nonetheless. Why do I miss it, I wonder, foul as it was?

I soon fell into a routine. I'd wake up on my own, go downstairs and pour myself cereal and eat it. I'd then get the newspaper, which was delivered to the front door, the Charleston *News and Courier*, lie down on the living room rug, and read the comics before school. I followed one comic strip in particular, *Dondi*. It was about a war orphan brought back to the United States. I read other comic strips each day, but I went to *Dondi* first.

Each morning I'd ride a Schwinn bike I'd picked out soon after we arrived in Charleston to the school Mom enrolled me in. I had no Jamie to walk to school with, but I had my red Schwinn, which I liked a lot. I liked being able to ride fast, no-handed, with my Army surplus backpack on my back.

In third grade, I attended my first private school. I didn't know the difference between private and public school. I didn't know any of the other kids at my school because Trippy went to a public school. I would later learn that my grandmother paid my tuition and all my other expenses, even though Uncle Unger could easily have afforded to.

After school, I'd bike home. Victoria would have made me lunch, which I would eat on my own, usually in front of a TV watching *As the World Turns*, since my mother and Uncle Unger ate "dinner," which the midday meal was called, later, around two o'clock. This meant he started drinking earlier than he used to in Chatham, as drinks preceded the two o'clock dinner. The evening meal, supper, usually found everyone so sloshed that Victoria or Georgiana, our other maid, or sometimes my mother, or sometimes I myself, would put together a meal for me. I started to like to cook.

In the afternoon, I would find random things to do, like put together a birdhouse from a kit someone had left in the shop/greenhouse abutting the main house. After putting it together, I hung it with string from a pecan tree in the yard and stocked it with birdseed bought at the Rexall drugstore a few blocks down King Street. I liked my birdhouse. I grew into a bit of a birdwatcher, prizing the bluejays and cardinals and other birds I didn't recognize who stopped by to eat.

I also joined the Cub Scouts and would go to meetings once a week and work on projects with other kids. I didn't make any good friends there—being a Yankee, I talked funny, so while the other kids were perfectly nice to me, they didn't go out of their way to befriend me.

One time a kid in my class invited me to his house for "dinner." I remember sitting at a rather formal dining room table eating with his family and thinking this was great, I'd made a new friend.

But not long after that he came up to me and told me that I had been a really nice kid at first, but I had changed, so we couldn't be friends any longer. This was actually OK with me, because I had my classmate, Bobby Hitt, as a friend, and I liked being by myself. But I wondered in what way I had changed enough that this kid went from asking me to his home for dinner to dropping me as his friend. Looking back now, I think it was either that his parents had decided he shouldn't be friends with a Yankee, or the wear and tear of living in the same house as Uncle Unger was taking its toll and I was indeed changing for the worse in ways I was not aware of.

What I did most on those afternoons was develop an ardent love of movies. There were three theaters within biking distance of our house: the Gloria, the Garden, and the luxurious Riviera. The Gloria and Garden cost ten cents admission, while the Riviera was twenty-five. Popcorn and a soda were a nickel. I went to the movies as much as I could. Nobody cared, which was great. For a kid, it was a dream come true. Do whatever you want.

I saw lots of war movies and westerns, with actors like John Wayne, Richard Widmark, Fess Parker, and Edward G. Robinson, whom I especially liked because his first name, Edward, was the same as mine. I liked the Three Stooges movies probably best of all, but I liked pretty much any movie that had a plot.

The Riviera showed the more sophisticated, adult movies like *Cat on a Hot Tin Roof, The Inn of the Sixth Happiness,* or *Separate Tables,* which I would go to because they were movies, but I thought they were boring. I didn't like them nearly as much as war movies, westerns, or comedies.

I remember laughing so hard, all by myself, that I wet my pants while watching *No Time for Sergeants* with Andy Griffith. The scene when he makes the toilet seats go up totally cracked me up. I also loved the epics. I went to *The Ten Commandments* seven or eight times, the same for *Ben-Hur.*

Movies meant the world to me. They gave me a new world, not so much an escape as an entrance into action, adventure, laughter, spectacle, all under control, with a beginning, a middle, and an end. Movies were paradise for me during those years, a godsend.

16.

Not long after we arrived in Charleston, and before I started school, I found myself in a church getting baptized. I think I was eight.

Other than Marnie teaching us about blackness at the end, and being told that Grandpa Kent had been a minister, I knew nothing about religion.

I did know about God, though. My mother would say the Lord's Prayer with me every night, and she would tell me never to worry about anything because "God is everywhere." That had been my religious training but it worked. I believed her.

There was also a Quaker theme that my Gammy Hallowell continued, I believe out of respect for our ancestor Lucretia Mott. Gammy considered herself a Quaker of a very vanilla, not devout, sort and went to Quaker Meeting now and then in Sandwich. She told me how proud she was of the Sandwich meeting because it was—is—the oldest continuing meeting in America, having started in 1656. I was impressed, but not for the reason Gammy would have hoped. I was so young I thought the "Sandwich meeting" was where they served sandwiches.

The most noticeable manifestation of Quakerism in my early childhood was that Gammy Hallowell, Duckie and Uncle Jimmy, and Lyndie and Jamie all called each other "thee" and used the possessive pronoun "thy." So when Duckie would greet me, she'd say, "Well, there thee is!" or "How is thee?" I grew so used to it that I didn't even notice it but when friends visited they'd always ask me what it was all about.

Dad thought it was stupid, so we never got into the thee-thy habit. I always kind of liked it and wished we had adopted it.

But no one—not Duckie and Uncle Jimmy, not Mom or Dad, and only rarely Gammy Hallowell—had gone to church or Quaker meetings regularly, or any other sort of religious ceremony. That I said the Lord's Prayer before I went to bed set me apart from Jamie and Lyndie, but we never talked about it. They didn't tease me about it or anything like that, and I really didn't make anything of it. Jamie and Lyndie said "thee" and I said the Lord's Prayer, but we didn't make any more of it than we did of preferring Whiting's milk to Hood's. In fact, that we all drank Whiting's mattered more, just as we all knew Fords were classier than Chevrolets.

But there I was, in a church, about to be baptized. My brother Ben joined the ceremony, as did my mother. Uncle Unger had been baptized years before. It turns out that joining the Episcopal Church was what well-bred white people did in Charleston.

So off we went to St. Michael's. Chance Ravenel, Trippy's mom, was my godmother, and Hugh Stiles, Uncle Unger's cousin, whom I called Uncle Hugh, was my godfather. Uncle Hugh was a doctor, the team doctor for the football squad at the Citadel, and one of the nicest men I ever met in my life.

DeWolfe Perry was the minister, or what Episcopalians call the rector. He dripped water onto my forehead, and the foreheads of the others getting baptized, and prayers got said, including the one I knew, the Lord's Prayer. Before I knew it, I was an Episcopalian.

I was signed up for the choir. Whoever did it didn't ask me, I was just told to go to choir practice at the church on Wednesdays at five o'clock. So I did.

The choir director was also the organist. She was a beautiful woman. I was always amazed that she played the organ not only with her hands but with her feet.

There was also a pretty girl in the choir named Tinka Perry, the rector's daughter. I never told her how pretty I thought she was, nor did I ever have a one-on-one conversation with her (until fifty years later), but she was my first crush.

Because of Tinka, but also because of the nice lady who directed us, I came to love choir. I would ride my red Schwinn to St. Michael's every Wednesday at five and ride it there again every Sunday morning at eight because Uncle Unger and Mom never went to church, opting to sleep in and recover from the night before. My mother would wake up to tie my tie, then fall back asleep while I got on my bike. Once at church, the choir members would don our red cassocks and white surplices, and we'd process, sing, recess, and have fun.

After services, there was always a wonderful breakfast. It was called "coffee hour" but there was much more than coffee. The highlight was the ham biscuits, "beaten biscuits," which meant the dough was beaten for fifteen to thirty minutes, until it blistered. This gave it a special texture and flavor able to stand up to the thin slice of salt-cured ham that would be laid within it, slathered with mustard butter. I could have eaten them all day, along with the copious amount of orange juice I washed them down with.

St. Michael's became my clubhouse and the choir my club. I listened to the services, and Reverend Perry's sermons, but they zipped right over my head. Many of the hymns stuck with me, like "Holy, Holy, Holy, Lord God Almighty," or "A Mighty Fortress Is Our God," or "Jesus Christ Is Risen Today," and, of course, all the Christmas carols.

The choir rehearsed for months for the big combined-choirs Christmas service. I felt I was part of something special, rehearsing for that event, one eye on Tinka, the other on everyone else, singing, "O Come, O Come, Emmanuel." I didn't know who Emmanuel was, but I joyfully sang and invited him to come, O come. I was really pretty clueless about theology and the various beliefs I supposedly held and affirmed every Sunday. I didn't know what the service was all about but I still loved church.

And not just for the ham biscuits, the singing, and Tinka. I loved the feelings I got at St. Michael's: the joy I felt whenever I went to church, the happiness and hope. I never left church or choir practice feeling anything but happy and glad. Those bike rides to and from St. Michael's were some of my happiest times in Charleston, right up there with the bike rides to and from the Garden, Gloria, and Riviera.

Sometimes I would take home with me the little pamphlets that were on display outside the church office. I didn't actually read them—they were about prayer and eternal life and the promises we make to God and such—but it wasn't the content I cared about so much as that they were mementos of church, keepsakes from a place I loved.

I didn't notice that no black people attended the church. I would learn many years later that after one of the services during my years in Charleston, one of the wardens (in Episcopal churches there is a vestry, a committee of people that run the church, which is headed by a senior warden and a junior warden) told Reverend Perry that he was going to report him to the bishop for allowing a black man to sit in the congregation one Sunday. If he ever did report him, nothing came of it. I would also learn that "colored people" were allowed to sit in the balcony for a wedding or a funeral of a person in whose home they worked, but otherwise they were not allowed in.

One day I built a little altar in my bedroom on the third floor of our new house on the harbor, right near the Battery, within sight of Fort Sumter, where the Civil War started. I had a window with a deep windowsill that looked right out onto Charleston Harbor.

I took some linen napkins from the drawer in the dining room and went up to the attic where some candlesticks were stored and set up an altar on the windowsill, which was more than deep enough to hold all my paraphernalia. I stocked my altar with the pamphlets I had taken from church along with the Book of Common Prayer I'd been given as a baptism present by Uncle Hugh and Chance.

I also found a red upholstered footstool in the attic that I used to kneel on in front of the altar. It was pretty similar to the stools we knelt on in church, so it felt authentic to me.

At random times I would kneel on the footstool, bow my head, and pray. I would always say the Lord's Prayer. Sometimes I'd read from one of the pamphlets or from the Book of Common Prayer. I didn't know what else to pray, so I would just ask for what I hoped for, usually that Uncle Unger could be less mean and that my mother could be happy.

I came to love my bedroom on the third floor of 76 East Bay St., the house we moved to from 75 King St. Not only did I have my little altar with its view of the harbor, I could also see the old rotting ship that Bobby Hitt and I turned into a fort/clubhouse, and I had my magic carpet of a TV. That bedroom became one of the many places I created in my life where I could feel safe and cozy, a place where I could live in fantasy as much as I pleased.

17.

Sober, Uncle Unger was a charming man, urbane, witty, a one-of-a-kind character, a man, even after I came to hate him, whom I emulated in many ways and who to this day makes up a significant part of who and what I am.

When the alcohol hit him, it changed a man who, when sober, looked rather like Clark Gable into a living gargoyle, a truly horrifying sight. It was usually gin that unleashed this poison, but sometimes it was bourbon. When he'd had more martinis than anyone could remember or count, a terrible, buried part of him rose up and slithered through him, taking possession of him altogether. The ominous telltale signs were the many pat phrases Uncle Unger used only when drunk, like the aforementioned "Shove off!"

I'd learned long before to disappear to my third-floor room when I saw what was coming. I'd curl up on the bed in front of my treasured black-and-white rabbit-eared 1958 television, my trusty escape hatch, as reliable as the Gloria, the Garden, and the Riviera were, where I would watch the shows that magically took me up and away; simple, straightforward dramas and a few comedies I will remember forever, like *You Bet Your Life*. (Groucho Marx was one of my earliest intellectual heroes; Bobby Hitt and I said if Groucho could debate Khrushchev, he could end the Cold War.) I'd watch TV until the yelling downstairs stopped or I fell asleep, whichever came first.

One night when I was about nine, it was not the usual crying I heard but a scream, sharper and more desperate than a yell. I hurried down the two flights of stairs to find Uncle Unger standing over

Mom in the living room, a fireplace poker held above him. She was on the floor, hand raised over her head to try to protect herself.

I ran at Uncle Unger as hard and fast as I could, tackled him around his waist, and pushed him away from my mother. He staggered but—in spite of how drunk he was—being the athlete he once had been, didn't fall over. He looked down at me and slurred, "You, want to join the party? Welcome aboard." ("Welcome aboard" was another of those annoying phrases that came out when he was drunk.) "We're just starting to have some fun down here. You could learn something."

I pushed him again, and he staggered over to his favorite armchair. After what looked like a moment's thought, he did a half pirouette, with one arm raised like a troubadour, and let himself fall into his chair. He easily could have hit me, or my mother, with the poker or the back of his hand, or anything, but he didn't. He never did. No matter how drunk, he never hit me. Must have been part of that code of honor. And as much as he threatened to beat my mother, I do not believe he ever did that, either.

I stood and stared at him in his chair as my mother got up off the floor and took a seat on the couch. "Come here," she said to me.

"I hate you," I told Uncle Unger. "I hate you."

"Good for you, boy," he said with a laugh. "You *should* hate me. I'm glad you have the gumption to say it. I watch you on the playground—you know, I can see you out there from the porch? And you never mix it up in the football games at recess, you're usually off with Bobby Hitt. I do *not* cotton to that boy, so I'm glad to see you can stick up for your mother like that. Never let anyone attack someone you love." More of the code, I guess.

"Don't you ever do that again," I said.

"Do what? Play with your mother? She's my wife, boy. I will do with her whatever I want to do."

The poker had fallen to the floor. After he told me he'd do whatever he wanted to do to my mother, I walked over to the poker and picked it up. I was actually just trying to get it out of the way but he must have thought I had a different purpose in mind because he said, with what I detected as actual fear in his voice, "Easy there, boy, easy. Everything's under control. The ship is steady as she goes." As I replaced the poker in its stand, I felt a thrill of triumph.

Then he seemed to start talking to himself. "The seagulls are circling the Yardarm, it's time to put some lines in the water and defrost the steaks."

The Yardarm was what he called the camp—a little shingled shack where you had to pump your water and use an outhouse—that he had on Monomoy Point off Chatham, where we used to go often before he moved us to Charleston.

I could tell he was close to passing out and that the danger was over. Before he fell asleep, I took advantage of his drunken, unguarded state and asked him what I'd always wanted to know: "Why did you move us to Charleston? Why did you take me away from Jamie and Lyndie and Duckie and Uncle Jimmy? Why couldn't we have stayed in Chatham? I hate Charleston."

"Because I wanted to get away from her people," he said, looking at my mother, "and come be with my people. Why not? I'm over sixty, I pay the bills, I'm captain of the ship, why shouldn't I live where I want? If you have any objections, just shove off."

18.

The years I was in Charleston, Uncle Unger drove a black Thunderbird. He kept it parked under a tree next to the house on East Bay St. Robert, an elderly black man who did tasks not suited to the maids Georgiana or Victoria, polished the car regularly.

Robert and I became friends. He was a smart man and I enjoyed talking with him. He made me laugh by taking out his false teeth and showing me his bare gums.

"Why did you lose your teeth?" I asked.

"Just you wait, boy, you be losin' yours one day, too." Robert was rather slight, had white curly hair cut very short, and always wore loose-fitting short-sleeved shirts. "What you doin' today, Mr. Ned? Mr. S. in a good mood?" he asked in a voice that was gravelly due to all the Camels he smoked. He always had a pack in his breast pocket. When he'd lean over to polish the car, I always worried they'd fall out, but they never did.

"I don't know," I said. "I don't like him. He's mean a lot of the time."

"I know," Robert said, "but don't you be tellin' him I said that. He must have a lot on his mind, that's all I can say. A man who has all he has, to be so disagreeable, don't make sense, if you ask me."

"Is he nice to you, Robert?"

"Boy, how long you live here? Not long, am I right? You from up north in Yankee territory, am I right?"

"Yes, Robert, I am from Chatham."

"Well, I don't know where Chatham is. Do they have any people my color up there?"

"Oh, yes, they do. Uncle Unger's maid was colored, and she and I and her friends were all friends."

"Well, you asked me if Mr. S. was nice to me. Let me ask you this. Have you ever seen a white man be nice to a colored man in this city?"

"I don't know, Robert. I guess I haven't noticed one way or the other."

Robert grinned and took out his teeth to make me laugh. "Well, don't you waste no time goin' 'round checkin', because I can tell you, you ain't gonna find no white man bein' nice to a colored man unless he be tippin' him for shinin' his shoe."

Robert started buffing the black T-bird with renewed vigor. "Take this car. Mr. S. love this more than he love any person of my color, that's for damn sure." Robert giggled at the idea. "That's for *damn* sure."

"Does he say thank you to you for polishing the car?"

"Whooo-wee, boy! What world you be livin' in? He might have done so when he lived up north, but down here, that's just not the way it's done. He's gotta fit in, I gotta fit in, we have our ways. I don' mind, though. I know my place. I know my place. Lotsa young coloreds be gettin' into all kinds of trouble these days by stirrin' things up, that's for sure. It's all gonna come crashin' down before we know what's happenin', I can promise you that. Now you don't tell Mr. S. what we been talkin' about or he fire me on the spot for sure. You promise me, Mr. Ned?"

"Robert, you shouldn't call me mister. I'm nine years old."

Robert kept buffing.

"Does he say thank you to you for anything?" I asked.

"Of course he does. He's a fine Southern gentleman, mighty fine Southern gentleman. They're all fine Southern gentlemen." Robert laughed. "Just don' be askin' him *why* he say thank you. Because then he have to tell you it's just another way of his sayin' 'I'm better than you, boy.'"

"I don't understand, Robert."

"Of course you don't, and I'm glad you don't. There's a whole lot you better off you not understand."

"What did you mean when you said it's gonna come crashing down, Robert?" I asked.

"I mean these young people ain't gonna sit still no more for what me and my people sat still for, that's all. And they ain't gonna care what price they gonna have to pay. I hope I live to see that day. But, then again, maybe not. It ain't gonna be pretty." He paused and looked up at the sky. "Robert, you better shut your mouth. Mr. Ned, you promise me, you don't say nothin' about what I been sayin'. Promise?"

"I promise," I said. "Will you promise not to call me mister?"

"You got a deal." Robert went back to buffing.

"How long are you gonna polish this car? It's really shiny now."

"I like to take my time. What else I gonna be doin'? And after I finish makin' this car shine bright as can be, then I start in on the brass plates on the doors. You ever see how they shine? That's because of me, Robert. I make things shine. I stay out of trouble." He stopped buffing for a moment and looked up. "Why am I talkin' to you so much? I will tell you why, Mr. Ned—I mean Ned. Because I like you!"

"I like you, too, Robert! We're friends, aren't we?" I put out my hand to shake.

Robert put down his buffing cloth, wiped his hands on his dark trousers, and shook my hand. As always, I did what I'd been taught to do and made good eye contact. For the first time I noticed how yellow the whites of his eyes were and how brown the pupils. I wondered if yellow meant he was sick, but I didn't know who to ask. If Uncle Hugh had been around, I would have asked him.

Robert was one of the people I missed after I left Charleston, not to return for forty years.

19.

One night Uncle Unger got me out of bed and said it was time to go for a ride. I had no idea what time it was, but this sounded like fun, so I got dressed and went downstairs where he waited impatiently.

"Your mother is asleep. I want to show you how fast this baby can go."

"OK. Why now?"

"Because, my boy, it feels like the right thing to do. Make a man out of you. I think we can have some fun. How about that?"

Uncle Unger stopped at the coatrack in the hallway, took a beige sweater off it, and put on one of his tweed driving caps. Outside, I got into the Thunderbird's passenger seat, a bucket seat upholstered in red leather, while Uncle Unger got into the driver's side.

Soon we were on the outskirts of Charleston, heading down a country road. "Why don't you get into the back seat and sit right behind me. You can be the navigator and keep an eye on the speedometer and tell me how fast we're going, OK?"

"OK." I crawled into the back seat, taking up a post right behind Uncle Unger. The round speedometer was framed by a cylindrical piece of black plastic. The numbers were white, the background black, and a red needle pointed to the numbers.

"Ready for takeoff?" he asked.

"Sure."

With a lurch, the T-bird picked up speed fast. "Read off the numbers, navigator," Uncle Unger said.

"Sixty," I read, "sixty-five, seventy, seventy-five, eighty."

"OK, good, steady as she goes. Now let's see what she's really got." With that, he pushed his sneakered right foot down on the gas pedal to the floor. I could see him push back in his seat as he did it.

The thrust threw me back but I recovered quickly. "Ninety-five!" I called out. "A hundred, a hundred five, a hundred ten!"

We were whizzing down a pretty narrow road. We were far out of town, so remote that there were no streetlights, only the moon and stars.

"No need to worry about cops out here." Uncle Unger laughed. "We're free as the air. Doesn't it feel grand?" I remember that word, "grand." It seemed like a strange word to use.

We screeched as we took a corner and I was thrown across the back seat against the wall on the other side. "I'm scared. Can we slow down?"

"You'll get used to it, boy," he said. "It's good for you. Make a man out of you. Call out the number, navigator."

"It's one twenty. Please, can we slow down?" The top speed listed on the speedometer was 140.

"Just keep reading the number," Uncle Unger said. "Do your job."

"It's one twenty-five now. Please slow down. *Please.*"

"Don't worry. You're in good hands with me."

I saw the keys in the ignition and thought of reaching forward between the front seats and turning off the car, but I didn't know if that was dangerous or not. So I just held on to the back of Uncle Unger's seat as tightly as I could. We flew past the landscape and the occasional house as the moonlight flitted between telephone poles and trees. The broken lines down the middle of the road blurred almost into a solid line, we were going so fast.

"We could crash," I said.

"That's what makes it fun. I know what I'm doing, boy. You just sit tight and enjoy the ride."

I don't know how long we kept driving at 120 mph through the twists and turns of the country roads outside Charleston, but I do know that I was so tense with terror the entire time that it wore me out. I don't know if it made a man out of me, but at least we survived to live another day.

When he finally slowed down, he said, "Good job, navigator. You can be my crew any day."

At the time I hadn't realized how drunk he was. I should have known because that's when he did crazy things. It also makes it even more amazing that we didn't crash.

He was a complicated man, full of charm, brilliance, style, and genuine goodness, along with the darker side that he brought into his marriage to my mother. I never got the full story about his first wife. I knew that her name was Anne, and that she was the mother of Allison and Hope.

In my imagination, I've always envisioned Unger and Anne as very much in love, but in the way the New York literary set does love: cynical, witty, hard-drinking, and without illusions. I imagined they laughed a lot, danced a lot, talked a lot, and shed few tears. I imagined they cast a cold eye on life, and on death.

I did all this imagining as I tried to put into context the one terrible event I know did happen, the awful thing that I believe (but certainly do not know) created Uncle Unger's darkness. One day Anne went into the kitchen in their house on Bridge Street, took a knife, and drew it across her neck. To finish the job, she walked, spurting blood, into the living room, found the pistol Uncle Unger kept in a drawer "for target practice," and shot herself in the head.

I didn't learn about this until after my mother had divorced Unger, and by then he and I had no relationship, so I was never able to ask him about it. Many years later I'd ask Hope why her mother had done what she did, and Hope said she honestly didn't know.

What a horrible thing for Unger and his daughters to have had to live with. I do know they were people who did not lead easy lives but who kept up their spirits. He had to have had a lot of good in him to have raised such lovely daughters.

Unger himself was not all darkness. I remember the night when I was the sickest I would ever be. At age nine, I got measles, and was covered with red dots. When nighttime came, I got hot.

My mother put a thermometer under my tongue, and when she took it out she called, "Unger, come here! Joseph, I need you now!" "Joseph" was a term of endearment they used for each other, dating back to when they went to see *Showboat* on Broadway and loved the song "Ol' Man River" sung by the character named Joe. After that musical, they started calling each other Joe or Joseph, at least when they were getting along.

Next thing I knew, Uncle Unger was in my bedroom. He almost never came up to my bedroom.

"His temp is a hundred and six." I could tell from my mother's tone of voice that 106 was not good. I could also tell I was not coming in right. I felt as if I was dreaming but I knew I was awake.

"Let's get him a cold washcloth," Uncle Unger said, "and call Hugh." I knew he meant Uncle Hugh.

A cold washcloth appeared on my forehead and it felt wonderful. I started to play with the curtain cord next to my bed. Mom had gone but Uncle Unger sat down on the bed and stroked my cheek with his hand. "You'll be fine, sailor," he said.

I liked his calling me sailor.

"Did I ever tell you about when I was in the Secret Service?" he asked, still stroking my cheek. Not waiting for an answer, he told me story after story that, in spite of my fever, my brain held on to, at least a couple of the details, which included "a snub-nosed revolver" and "some German spies" and "an old warehouse." The rest of the stories disappeared into my fever, but Uncle Unger did not leave my bedside.

Finally my mother came back and said to Uncle Unger, "Hugh's coming over with a suppository."

Within minutes, Uncle Unger was turning me over, lowering my pajama bottoms, and putting something up my butt. It felt strange but I didn't mind. Then he pulled up my pajamas, turned me over, asked my mother to put more cold water on the washcloth, and went back to stroking my cheek.

I heard Uncle Hugh talking to my mother across the room. I heard the words "very serious" and "we'll know by morning." But I was not in the least worried. I was in dreamland with a kind and protective Uncle Unger watching over me.

20.

On certain Sundays, after I got back from church, and after Mom and Uncle Unger got up and dressed, we'd all go to the apartment of Trippy's grandparents. They were the parents of J. J., Trippy's dad. Gramps and Gramps's wife, Mary Francis, lived in the upper half of a yellow two-family house across town. Like Trippy's family, they didn't have much money, but they had enough money and even more tradition to have relatives over for TV and meals. I liked going, mainly because the food was so good.

I never understood exactly how Unger was related to them, but he was close enough that we had many delicious Sunday dinners there: roast beef, ham, sweet potatoes with marshmallow, biscuits with butter and honey, grits or mashed potatoes and sometimes both, fried chicken, green beans with bacon, stewed okra and tomatoes, black-eyed peas, pecan or apple pie à la mode. I loved everything, except I hated sweet potatoes, even with marshmallow. Unger would force me to eat them anyway, which I thought unfair because when we had carrots, he didn't have to eat them, simply because he didn't like them. "When you get to be my age, then you can decide," he'd say.

Sometimes Oral Roberts would be on the TV in a small sitting room off the living room. An ancient lady in a wheelchair would always be there watching that TV, sitting just about three feet away from it. They called her Aunt Bessie, but she didn't eat with us, no one ever talked to her, and she didn't speak to anyone.

Until Charleston, I'd never seen anyone or anything like Oral Roberts, or Aunt Bessie for that matter. I would watch him on the

TV while sitting on the floor next to Aunt Bessie, and I'd feel increasingly uneasy as he'd get worked up about casting evil spirits out of the sick and crippled people who'd hobble or sometimes even crawl up to him at his altar to be healed. One at a time they'd come up out of an audience of thousands and kneel on a platform in front of the Reverend Roberts.

Roberts, his dark hair oiled straight back and his flashing eyes ablaze, would put both his hands on the head of a woman kneeling before him. Then he would close his own eyes tightly, raise his head up toward the roof and heaven beyond, and mightily bellow, "Dear *Jeee*sus, sweet *Jeee*sus, hear me, Jesus, show her your divine mercy and with your almighty power cast out these cursed spirits, cast them out of her, cast them out!"

The woman would rise up, miraculously healed, and walk a few steps with no limp at all. I'm surprised she didn't dance a jig on the spot. Instead, she'd hold her hands to her face, weep, and fall to the floor in tears of joy and religious transport, before getting up again and allowing another to kneel and be healed.

This upset and frightened me. Could Oral Roberts really have this power of healing? On cue, on Sunday mornings? If so, why did people go to doctors? I suspected it was fake, but at the same time I wondered if it might be true. His way of laying on hands and calling upon Jesus totally creeped me out, though. It was *not* what I was used to. On the other hand, when I told our maid Victoria about it, she said, "Oh, child, I watches Reverend Roberts every Sunday. I never miss it. He has the true gift of healing." So I tried to get used to him, but whenever I watched his act I felt that something wrong was being done in the name of something good.

It was confusing. I took such solace at St. Michael's Church and the community there, found comfort and actual fun in building the altar in my bedroom and in praying daily. I felt that God was my friend, but at the same time I was deeply unsettled by the religious fervor of Oral Roberts. How could religion on the one hand be such a force of hope and stability in my life, but when expressed with a certain kind of intensity and fervor become utterly destabilizing, frightening, and alien?

Years later, when sitting in the company of psychotic people, I'd get a clue. Any emotion ramped up high enough becomes destabilizing, or, as my teachers would put it, "Psychosis is the mind's last defense against unbearable affect." At the right temperature, love may create the best feeling in the world, but when it gets too hot it can make us mad.

Of course, I had no one to guide me along these lines in Charleston. I just had my own wits, and the angels who've always seemed to have my back.

Not just on Sundays, but sometimes we'd also go to Gramps and Mary Francis's place for supper on Friday nights and watch the Gillette *Friday Night Fights*. I remember seeing the great Sugar Ray Robinson, but what I remember even better than the boxing is the jingle Gillette made so popular: "If you want to look sharp, and feel sharp too, get the razor that is built for you . . ."

Gramps would sit in the wingback chair that amounted to his throne and mumble comments on the proceedings. He was a hide-bound racist in the tradition of Strom Thurmond, but this was when I was only just learning what a racist was. When he'd talk about niggers (I don't remember the first time I heard that word, but it was in Charleston) and the world going to hell and perdition (certainly didn't know what that word meant) because it was only a matter of time before the races started to mix, I felt completely confused. Didn't we already mix? Colored people lived with me and were instrumental in raising me. I didn't understand what Gramps had against them. As for Gramps, all I knew about him was that he was old, quite rotund, bent over when he walked, and very tired all the time.

His wife, Mary Francis, on the other hand, was just about the peppiest and sweetest lady I ever met in Charleston, even though one day she told me I deserved "to be taken into the backyard and given a sound thrashing" for telling Uncle Unger that I hated him after he told my mother to "shove off."

Some Sundays, we would go to a restaurant called Henry's on Market Street, right across from the roofed-over set of public stands and stalls where people sold fresh vegetables, baskets, shrimp, and all kinds of other goods during the week. It was closed on Sundays, though, and would be empty except for kids riding their bikes around and through it. Sometimes I'd see white boys riding by calling out the black boys. But I never saw any fights. All that wouldn't start until a few years later, just as Robert predicted.

We'd sit in a booth, my mother and Unger next to each other. They'd usually order Manhattans and give me the cherries. After a while I'd get bored watching them sip their cocktails, so I'd go out and ride around on my bike in the deserted market, imagining I was a detective or a superhero in search of bad guys. Then I'd go back in to eat.

When I finally went back in—I knew to take my time because Unger liked to savor the two Manhattans he had first—our dinner had arrived. Unger was having planked steak, rare. Planked steak was his favorite, which I thought was really cool, first because the steak came on an actual oval-shaped board, and second because it was surrounded by a thick, browned coil of mashed potatoes. Unger enjoyed the fact that I liked it, and we would happily commune on the merits of planked steak.

He always wore a sports jacket, usually tweed, sometimes camel-hair; one of his Brooks Brothers button-down shirts, usually yellow, white, or pink; and a carefully knotted tie set off by a gold collar pin in the shape of a classic safety pin. Even though he knew what he was going to order, he'd study the menu as if looking at it for the first time, wearing his thick tortoiseshell reading glasses, which he would remove after ordering but then don once again to study the check at the end of the meal—and I mean study. The bill could not have been complicated, but he would bear down on it as if Henry's restaurant and he were in a competition to see who could cheat whom.

Sometimes he'd get mad at me because I'd inadvertently kick his legs under the table and he'd order me outside. And sometimes I'd kick his legs on purpose, wanting to annoy him. This is how we were with each other. We fought every day and I hated him, but at the same time I was drawn to him.

I loved his collar pin—why does a nine-year-old boy love his stepfather's collar pin?—and I loved how his shoes were always shined, loafers the color and sheen of polished mahogany. He taught me to shine shoes, using a small brush to apply Kiwi shoe polish from a round tin you had to pry open with a coin, buffing them with a horsehair brush first, then with a shoeshine rag, adding a little spit at the end to make it "shine up real good. " He could be like a real dad. Without meaning to, he taught me that a kid can love someone he also hates.

It was all so confusing. One night Uncle Unger came home late from the annual meeting of the Yacht Club, which was only a few hundred yards from our house. It had been a night devoted to heavy drinking and gambling. Even though I was on the third floor, he woke me up when he came in with a couple of other men. The next thing I knew he was calling me down from my bedroom.

I got out of bed, went downstairs, and found him in his bedroom, my mother lying, with her nightgown hiked up, passed out on their bed, and two other men standing next to Unger shooting dice using my mother's uncovered, naked rear as the backstop. A big heap of cash lay on the bed next to her like so much trash.

"Watch," Unger said. "You might learn something."

I wasn't used to seeing my mother naked, even partially naked as she was then. I remember wondering if it was really her, even though I knew it was. I didn't know what craps was, so I had no idea why they were throwing dice against her, but I did know it was wrong. I wanted to stop them, but I didn't know how. Later I thought that I should have done something and felt bad that I hadn't.

Unger and his friends were so drunk they didn't notice me silently retreating upstairs to my room.

Nothing was said the next day. It was as if it never happened. I don't think my mother even knew it did. I didn't say anything because I was half-pretending it hadn't happened myself. I was long past the stage of lodging complaints.

It was always at night. Later that same year, Unger called me down to their bedroom in the wee hours and handed me a hatchet he'd given me for my tenth birthday. My mother was sitting in the loveseat across the room. "Boy, I want you to throw this hatchet so it sticks into the door."

"Why?" I asked him.

"Because you've never used this hatchet since I gave it to you and I want you to learn how." He didn't know it, but I'd been throwing that hatchet at trees outside and in Bobby's and my fort down at the old pier ever since he'd given it to me. I would just always return it to the same place, so he must have thought I never played with it.

I thought throwing a hatchet indoors would be fun. I would have thought it was against the rules, but I assumed Unger was so drunk he was suspending those rules.

But then came the catch. At night, there was always a catch. When he was drunk, that's when stuff got weird. "If the hatchet doesn't stick in the door and stay, then I'm going to pick it up and throw it at your mother's head."

This time he'd lost his power. Unlike the time when the men were shooting dice and I felt transfixed and scared, this time a switch got flipped. Instead of being terrified, I flat-out didn't believe him. He was trying to scare me, or toughen me up, or whatever twisted purpose he had in mind, I didn't know.

What bizarre and malign design he had in mind for me during those years with him I would never figure out, but I do believe a dangerous madness was in play over and beyond the effects of alcohol. There was something blazingly insane raging inside him. He hid it from the world, but not from my mother and me. It kept us on constant alert. But that night, his spell over me finally broke. Sooner or later, fear peters out.

I was able to manage his madness that night with no damage done. I knew he wouldn't throw that hatchet at my mother's head. I remember, maybe for the first time ever, not feeling the least bit afraid of him. He was just nuts. Still, I didn't want to call his bluff, on the off chance that as drunk as he was, he might throw it just out of drunkenness.

"You better hit that door," he said as menacingly as he could manage—an Edward G. Robinson snarl if ever there was one.

I calmly took the hatchet from him. Because I'd thrown it at so many trees, I knew it would stick. I wanted this charade to be over, so I didn't give Unger the drawn-out drama he was looking for.

I just threw the stupid hatchet. It stuck in the middle of the bedroom door with a shudder. Perfect hit.

He was disappointed. "Go back to bed," he said gruffly. I looked over at my mother, whose eyes were closed. I was glad that she'd passed out. I knew she was safe. And for the first time in the longest while, I knew that I was safe as well.

21.

Not long after that night, my mother asked me if I wanted to go to boarding school. "What's boarding school?" I asked.

My mother laughed and explained. "I've talked to a friend who lives in Chatham in the summer. He runs a school just outside of Boston. His name is Hart Fessenden and his school is called the Fessenden School. He said you could start there in September if you want to."

"So, what would happen to you and Uncle Unger?"

"We'd stay here, and you'd live up there. On vacations you could go stay with Jamie."

I could see she had tears in her eyes. "Why are you crying?"

"Because this is the hardest thing I've ever done, but I know it's the right thing, as long as you want to go."

"Why is it hard?"

"Because I will miss you so much," she said, dabbing her eyes with a tissue. "But this is not a good place for you to be anymore, and I think we both know that."

My mind was racing. "You mean, I can go live at a school and spend vacations with Jamie? Wow! You'd let me do that?"

At last, my mother smiled. "Let you? I *want* you to do that!"

Shopping for Fessenden required me to acquaint myself with clothes I'd never worn before but would wear for the rest of my life. Brooks Brothers button-down shirts, rep ties, khaki trousers, blazers, penny loafers. The school also actually required that I buy a fedora. Thank God I never had to wear it. A ten-year-old in a fedora beggars belief. In any case, I was baptized into the preppy attire I would wear forever after.

Lyn, Duckie, and Mom dropped me off in September 1960. I'd never seen a school like it. Ivy-covered buildings surrounded by playing fields. After we parked, we took my trunk and looked for directions to my dorm.

Walking the grounds, Duckie suddenly said, "Look over there. That's Bette Davis."

"Wow, I think you're right," Lyn said.

"Who's Bette Davis?" I asked. It actually was Bette Davis. Her son Michael attended Fessenden for a while.

A man approached and shook hands with us all. "I'm Charlie Guss," he said to Duckie, my mother, and Lyn. "What's your name?" he asked me.

"Ned Hallowell."

"Well, Ned Hallowell, welcome to Fessenden."

Like almost all the teachers at Fessenden, Mr. Guss was really nice. He was balding, had a mustache, and stood about five foot ten. He taught English, and all the kids called him Minny Goo. No one ever explained to me how that name came to be.

Mr. Guss walked us up to my dorm. My room was a cubicle just big enough for a twin bed, a bureau, a small desk and chair, and a closet created by a curtain. The door to the cubicle was a curtain with green and white stripes, which I can see now as vividly as in 1960. Unpacking my trunk and putting shirts in the bureau, I encountered the smell of unpainted white pine furniture that I will always associate with Fessenden.

We soon met Mr. Brown and Mr. Cook, the dorm parents. Although Mr. Brown could be a real terror, both Cook and Brown were basically good guys.

I hugged and kissed Mom, Duckie, and Lyn goodbye, then sat down on my bed. Out the window was a pretty garden. I heard other kids and other families settling in. I bounced up and down on the bed a few times and checked to see if I felt homesick, as Lyn had told me I might. Jamie had also given me advice not to use our family's words for private parts and bodily waste. I did not feel homesick, and resolved to heed Jamie's advice.

I was ten years old, embarking on a long stint in boarding schools. For some kids, this might have felt frightening, even tear-jerking. For me, it felt exhilarating.

I took to the school right away. None of the teachers were drunks, at least as far as I could tell. There was a predictable schedule to the day. People were genuinely nice. None of the adults fought with each other.

From day one, I excelled academically. For the first time in my life, I studied. We had study halls twice a day, before and after dinner. With nothing else to do but study during those hours, I applied myself. But I would have tried hard anyway, as I wanted the teachers, the kids, and the school to like me. I had longed to find a safe haven at Fessenden, and I did.

The four years I boarded there were happy during the daytime, sad at night, as I would go to sleep worrying about my mother. Before bed I'd say a prayer, asking God to make her happy. But during the day, Fessy, as we called it, was fun. The food was delicious. They had two cooks, chefs really, named Peggy and Howard. We kids didn't know how good we had it. I'd stack the food at Fessy up against institutional food anywhere. I can still almost taste certain dishes. They made a simple dessert called cottage pudding that wasn't pudding at all but rather a yellow cake you poured a delectable caramel sauce over. And there was a lunch dish called cheese dreams that I'd pay dearly to get the recipe for today. It was a cheese concoction that included onions, peppers, and various seasonings atop a toasted English muffin. I don't know what its secret ingredients were, but if they had let me back then, I could have eaten a dozen cheese dreams at every sitting.

Most people talk about how they hated grammar, but the way Mr. Cook, Mr. Saunders, and Mr. Gibson taught it made it fun. I'll never forget diagramming the Lord's Prayer and "The Star-Spangled Banner." Those teachers took a traditionally dry and boring subject and turned it into an inviting challenge. Fessy fed my love affair with words and the composition of sentences.

Mr. Slocum spent an entire term in eighth grade American History having us study the Constitution. We read the whole thing, accompanied by a cheat sheet that a brilliant local attorney had written up specifically for Mr. Slocum and us eighth graders. Only because of that course do phrases like "due process," "habeas corpus," "res ipsa loquitur," or "cruel and unusual punishment" have resonant meaning for me.

There was, however, a secret, terrible world at Fessy I knew nothing of. Completely unbeknownst to me, a faculty member at the time was part of a ring of pedophiles. He was arrested a few years after I graduated. I never heard about it when I was a student there. It saddens me very much that the place I found to be so safe, indeed the place that in many ways saved my life, turned out to be so horribly unsafe for other students, if not ruining their lives, then traumatizing and scarring them deeply.

But for me, Fessenden was a godsend. Fessy bailed me out of a terrible situation in Charleston. Thank God my mother was strong enough to send me away, and thank God my grandmother could pay for it.

22.

One day I got a note instructing me to report after lunch to the office of Dr. Merritt, the school psychologist.

I walked down the basement corridor, a corridor I'd sometimes run through in the middle of the night as a game, following the night watchman's rounds. I passed a couple of classrooms before arriving at the tiny piano practice room that served as Dr. Merritt's office. There was only room for an upright piano and bench, and two wooden chairs facing each other in front of a window.

When I arrived, the door was open. Dr. Merritt waved me in. I took a seat across from him.

Dr. Merritt wasn't young; he was quite puffy and chubby, dressed in a rumpled black suit, a wrinkled white shirt, and a nondescript black tie. He wore glasses, but one eye was covered with a white gauze patch. "I had a procedure on my eye," were his first words.

"Oh," I said.

Then there was silence. I had never seen a psychologist before, or any kind of professional in the world of mental health, so I didn't know what was up. I looked at the floor, waiting for some guidance.

"You're Hallowell?"

"Yes, that's me," I said. I didn't know if I should stand and shake his hand but decided against it.

"Your mother called the school and asked that you have a visit with me. Do you know why she did that?"

"No."

"Do you know who I am?"

"I think so."

"I'm the school psychologist."

"Yes, that's what I thought."

"Well, how about if you tell me about your life so far?"

I remember starting to talk and out of the blue the floodgates opened. I talked and talked and cried and cried. I do not remember a single word I said, but I was in that room for a good half hour. Dr. Merritt sat there, not saying a word.

What happened next makes me believe he was either the best or the worst psychologist on the planet. He said, "You can go now. We do not need to meet again."

I left his office, went to the gym, changed my clothes, and went out onto the soccer fields. I thought no more about my visit with Dr. Merritt.

Looking back, I would not have dismissed a ten-year-old boy who'd just tearfully sobbed out a story about family turmoil. But that's me. I've always been a sucker for a sob story. I've always been the one who jumps in to save. Maybe Dr. Merritt was smarter than that. Maybe he thought it best for me to seal over those memories and let Fessenden, rather than him, save me. And that's pretty much what happened.

23.

The varsity soccer field at Fessenden, adjacent to a grove of elm trees and the Swedish Home for the Elderly, holds two of my most haunting memories.

The first happened on a rainy day in September 1963. I was in the eighth grade, thirteen years old. We were practicing soccer in spite of the rain. I liked playing in the rain because I hate the heat and it kept me cool. In spite of never being much of a jock I had made the varsity, and I was lucky to have Harry Boyadjian as a coach. Mr. Boyadjian had come over from Jordan the same year I arrived at Fessenden, both of us to escape fighting, me the fighting in my home, and Mr. B. the mortar shells exploding on the grounds of the school where he had taught.

He emphasized a new concept to us American soccer players: *Play your position.* Don't just run around like undisciplined fools chasing the ball all over the field, or as Mr. B. called it, the pitch. This concept worked so well that Mr. Boyadjian's teams had been undefeated since his arrival at Fessenden.

On this particular day, as I was trying to focus on staying in my lane at what was called, back then, left halfback, I happened to look over toward the grove of trees, green and glistening in the rain.

I saw a man standing beneath the branches of one of the trees, maybe a hundred yards away. He wore a tan trench coat but no hat. I went back to focusing on practice, but something made me look over once more at that man under the trees.

With this second look, I saw who he was. It was my father. Why hadn't Dad told me he was coming? Why was he standing so far away? Why didn't he have on a rain hat or carry an umbrella? Why wasn't he at his school, teaching?

Mr. B. yelled at me to get back in the game and stay in my lane, so I waved OK and refocused on practice. The rain started coming down harder and the field became mushier by the minute. I wiped water off my face and stole a glance in the direction of my dad but he was no longer there.

To this day I have no idea why he appeared on a rainy day, not even a game day for us, stood by himself in the rain, a hundred yards away, never coming over to say hi or get a good look at our practice, and then simply left. I'd always wished he would come see me play.

That image stayed with me, my father standing so far away in the rain, wearing a trench coat with no hat, watching for a few minutes, not waving, not calling out.

The second image came two months after that. This time it was a sunny day, and we were playing a game against another school. Sometime into the first half the ref blew his whistle and announced, "This game is suspended. President Kennedy has been shot."

What? Was it possible? By the time we all knew that it was indeed true, we were walking back to the locker room to a world that had totally changed.

I felt as if I knew President Kennedy personally. And like so many others, I loved him. He was so funny and witty and able to make me and all of us feel good about life. I didn't understand politics, but as long as President Kennedy was in charge, I believed that everything would be just fine.

And then, out of nowhere, with no warning, someone had killed him. I was stunned, we all were. A few muttered about how they didn't like the Kennedys anyway, but that was a muffled mutter. The entire country lapsed into a numbed state of incredulity, even staunch

Republicans, of whom there were many at Fessenden. It was too much to believe all at once. For me, in many ways, it still is, more than fifty years later.

Dylan Thomas wrote, "After the first death, there is no other." For many of my generation, JFK was our first death, our first encounter with sudden, catastrophic loss that would leave us innocent no more, yearning for a time we once had but never would again.

24.

After seven years of marriage, my mother divorced Uncle Unger. My final contact with him was not actually with him, but with his hats.

That year of the divorce the three of us were summering in Chatham, living in the ivy-covered house on Minister's Point where my mother had spent much of her childhood. At one point, Grandpa Kent and two other ministers had owned all of what naturally enough came to be called Minister's Point. They'd bought it in the early 1900s for nine hundred dollars. Grandpa Kent soon bought out his partners and built the house covered with the ivy he'd started with a clipping he brought over from England. Next to that he built a windmill upon which he'd hang eels he caught, before skinning them.

Attached to the roof of the house was a quarterboard with the words LIGHT OF THE EAST engraved in it. After the house left our family's hands, we salvaged that quarterboard because it meant so much to my mother, but it has since disappeared.

The year before, Unger had told my mother he'd buy her the ivy-covered house because she loved it so much. She was beyond overjoyed; for her it was a dream come true. But then Unger said, "You didn't think I was serious, did you? I wouldn't any more buy you that house than pay for Neddy's tuition. I gotcha going though, didn't I?"

They lasted another ten months. When Mom told me about the divorce, I let out a whoop of delight and gave her a hug. "We have to go back to pick up some clothes," she said. "He's not there. He's out of town for a few days."

All I remember from walking into that little house that day was the rack of hats near the front door: mostly gray felt fedoras with black bands, one brown fedora with a gray band, a few flat hats, and one straw hat.

As Mom went around the house picking up what was left of her belongings and packing them into suitcases, I went at the hats. I started slowly but then built up into a frenzy I didn't even know I had in me. One by one I took each hat, ripped at it, pulled the band, gouged at the rim inside with its gold monogrammed ULS and then stomped on the hat until it was as disfigured as possible.

Never before or since have I flown into a rage like that. I would say it was uncontrollable, but that wouldn't be true. It was supremely controlled. I am quite certain that if Unger had been there, I would have done whatever I could do to kill him. Such is the primal nature of rage that there's an off chance I would have succeeded. Years later, when one of my patients ripped all the sinks and toilets off walls and floors of the bathroom on the inpatient unit, I understood where his burst of superhuman strength came from.

In those few minutes, I let it all out. When finished, I stood panting, looking down at all the hats strewn across the floor, hatbands lying here and there like jilted lovers after a lost weekend.

My mother walked through the mess on her way out the door, two suitcases in hand. I don't think she even saw the hats. There was so much she hadn't seen along the way, which was just as well.

I'd seen too much myself, which is why I exploded, taking it all out on his hats. Looking back, I wish Unger had gotten the help he needed. After his wife's suicide, rather than seek help for only his daughters, I wish he'd sought help for the person who needed it most: himself.

When he was on his game, he could be fun, witty, affectionate, and a good man. He could have taught me how to play baseball, a game he was skilled at; he could have taught me how to play bridge. (The story was, he was so good at bridge that he'd made his fortune gambling in high-end games on Atlantic crossings in the big ocean liners.) He could have taught me what he knew of the world.

Instead, with no one to stop him or help him, he let his demons take him over, retreating to Charleston to hole up, in the same way I think Uncle Jimmy retreated to Chatham. I like to think that with the right help he could have been a good stepfather, just as I like to think that with the right help, Uncle Jimmy could have become a success in business, and my dad would never have divorced my mother.

After I finished taking my fury out on the hats, I looked around. Minister's Point had given my mother so many happy memories, as well as me and our entire family. It was better to dwell on that. As I calmed down and took in my surroundings, I realized that Unger had always been an intruder. If I had been older I would have known that what he wanted from my mother she couldn't give him, any more than he could give her what she wanted. They had each lost their true love, one to death, the other to madness, and nothing could bring them back. It was a mismatch from the start: two broken people looking for repair.

25.

Three houses stood on the property we owned on Minister's Point. About fifty yards to the right of the ivy-covered house stood what we called the Big House, because, of course, it was so big. It was up a slope from the ivy house, atop a steep dune, with at least a fifty-foot drop to the beach, looking straight out to the bay, and beyond that to what we called the Outside Beach, where we often picnicked, then beyond to the Atlantic Ocean which we swam in during those picnics. The dune had a view that grown-ups marveled at.

What I loved was the roofed porch that ran almost all the way around the Big House. As kids, we loved to run up and down on that porch, hopping in and out of the many wicker rocking chairs the grown-ups would sit in at cocktail time while marveling at a spectacular sunset.

I can see Uncle Jimmy in his rocker, can of warm Pabst in his hand (one of his many eccentricities, he liked his beer at room temperature), Duckie with her bourbon and water on the rocks, same for Mom. They'd both be wearing summer dresses, while Skipper would have on his suit, white shirt, maroon knit tie, and boater as he rocked and drank his highball of scotch and water. Gammy McKey would have on her navy blue dress with white polka dots as she drank her sherry. It could get awkward if Aunt Ruth (not a real aunt, just one of the many close friends who got designated Aunt or Uncle) was there, as Skipper and she might have had an affair. She'd made a living playing the piano around the world, and also playing backgammon for money. Other random visitors would come and go, people whom all of us kids had to greet with a handshake before going on our way.

I loved the Big House because it was so airy and full of rooms with high ceilings, so high I don't remember what the ceilings actually looked like. There must have been fifteen bedrooms on its three floors, if you include the attic (and why not?). You could get lost in that house. It was so vast it echoed. But at night it wasn't scary because there were always so many lights on and usually a fire in one of the fireplaces, plus there were always lots of people there. No one ever stayed in the Big House alone, but no one family resided there all the time, either. It was a come-and-go house. I never knew which bed I'd be sleeping in until I got into it.

On the other side of the Big House, maybe a hundred yards to the right and down a slope, there was what we called the Little House, which actually wasn't all that little. Spillover from the Big House would stay there. I remember Aunt Madge, freckled, friendly, red-headed Aunt Madge from Nantucket, staying in the Little House one entire summer, and when we asked her one day, as we were driving somewhere in her car, if we could visit Nantucket, she said she'd love for us to come but—and these were her exact words—"I doubt it."

Lyndie and Jamie learned to swim on the beach in front of the ivy-covered house, in bathing suits someone—maybe Aunt Nell—had knitted for them. There are old photographs of Benjie and Johnny and Lyndie and Jamie all playing on that beach, along with Duckie and Mom and Uncle Jimmy. Dad wasn't in any of the photos.

Mom loved that ivy-covered house. She thought of it as her home. If only my family, under Uncle Jimmy's direction, had not sold Minister's Point for a song, we'd be wealthy today. Of course, we had many, many chances to become rich that we didn't take advantage of, mostly, I think, because our minds, the grown-ups' minds, didn't have the knack for amassing a fortune. It was like a missing gene.

Gammy Hallowell bought Ryder's Cove so my dad could get a job there when he got out of the mental hospital. He could have gotten a job anyway—he was a genius around boats—but she wanted to make sure, so she bought it for pennies on the dollar. Had she not sold Ryder's Cove when Mom and Dad got divorced, we'd also be rich today. Had we held on to just about any of the real estate we owned when I was growing up . . .

But making a lot of money was just not in our skill set. After Uncle Jimmy made the killing by investing in IBM way back, the Hallowells' moneymaking came to a screeching halt.

When my mother got divorced from Unger, I asked, "What will we do for money?" At least I had a shred of practicality in my fourteen-year-old mind.

"Oh, don't worry about it," she said, "we'll be fine."

My mother lived out her final years on a combination of what Ben, John, and I could send her, what Duckie and Uncle Jimmy gave her, and public assistance. Still, I don't think it was poverty that made her sad sometimes in the years before she died. It was loss. I don't believe either she or my dad ever got over losing each other.

26.

n the spring of my seventh grade year at Fessenden, my mother decided to take the blue Mustang she'd bought with some of the settlement money she'd received from Unger and drive up from Chatham to spend a weekend in Boston. She almost never traveled anywhere on her own, but Lyn wanted her to come up to talk about men—Lyn valued my mother's input on this one subject, the subject Mom knew better than any other subject even though that knowledge never helped her find a marriage that worked.

She checked in at the Howard Johnson's in Kenmore Square. I got a pass to leave school for the weekend to stay with her. On the Saturday afternoon she took me to the one and only Red Sox game the two of us ever went to together. I had fallen completely in love with the Red Sox, a terrible team, so going to a game was a huge deal.

Since Skipper was an avid Red Sox fan, my mother, like most of New England, had rooting for the Red Sox in her genes. Really, we all grew up loving the Red Sox except Johnny and Jamie, but even those two did not denigrate our fandom.

Of my two brothers, Johnny was the brain and Benjie the jock. Actually, Ben was just as smart as, if not smarter than, Johnny, but he didn't make academic achievement as much of a priority as Johnny did. Benjie loved sports. He was an All-Star catcher in the Cape League, and he would have been the starting catcher at the Naval Academy had not the all-world Joe Bellino been on that same team.

Skipper and Uncle Jimmy used to sit in lawn chairs on the hill overlooking center field at the Chatham baseball park, Skipper wearing his dark suit, white shirt, knit tie, and white straw boater, sipping his highball, and Uncle Jimmy in his sneakers and khaki trousers drinking his Pabst, while they watched Benjie play. Sometimes I'd come, too. After Skipper died and Benjie went off to the Naval Academy, Uncle Jimmy would still watch the town team, only now from his Jeep sipping tea, having given up alcohol.

Since I was not brought up with any religion until I got to Charleston, sports, especially baseball, took the place that religion occupied for others. Be faithful. Play hard but play fair. Never give up. Root for the underdog. Be a good sport. I would hear all these instructions and many more of their ilk while growing up.

Having been to a game only with my father, I didn't know what to expect with Mom. I appreciated her taking me, though. It was fun to do something together, and really fun to go to Fenway Park.

I wish I could convey now how beautiful she looked that day. If I could describe my mother's beauty, not an easy thing for a son to do, I'd say she was demure, like Deborah Kerr or Audrey Hepburn.

In spite of all her dissolute years with Unger, she still occupied her space in this world with uncommon dignity and grace. She could enter the finest club or the most exclusive gathering and be right at home. Even when drunk—a word that she'd never use— she was never sloppy or crude. Taught by Skipper and her teachers at Miss May's School, she embodied the best of the tradition of manners, gentility, and kindness.

The day of our game, she wore powder-blue slacks, a plain white blouse, and a pink cardigan sweater with large straw buttons. She wore the pearl tiara pin she always wore, a present from my father, I believe, and she had on her usual pearl earrings. Dressing for Fenway, she wore blue flats and an old floppy hat with a purple band. I am sure she also had on Arpège perfume, an aroma that activates her in my memory more instantly than any other cue.

We walked the quarter mile from the hotel to Fenway Park, where she bought two of the best box seats in the place for three dollars each. Attendance that day was 7,412 in a park with a 35,000 capacity. Getting great seats just walking up minutes before the game was no problem.

Knowing Fenway Park, I led the way, down the concrete ramp from the street-level ticket booths to the concession stands stationed cozily behind the third base line where our seats were. The most exciting plays happened at third. Also, from here I could see into the Red Sox dugout, blocked from view if you sat behind first base.

As we walked through the stands, I did my usual inner genuflect, marveling at how perfect a setting this place was for something I loved to do: watch baseball. Four rows behind the third base dugout, I easily found our seats. We had arrived early, so some of the visiting Detroit Tigers were playing catch in front of us, while Red Sox players were doing the same across the way. Other players did sprints in the outfield.

I went back behind the stands, part of my ritual, to get my mother a Coke and me two hot dogs and a Coke. I think the hot dogs went for 25 cents, and they tasted better at Fenway than anywhere else. I smeared the hot dogs with *tons* of mustard and packed them into a cardboard tray along with the Cokes and napkins (my mother would insist), making my way back to our seats lickety-split.

A teacher at Fessenden once asked me what my idea of heaven was and I immediately replied, "Sitting in a box seat along the third base line eating a hot dog in a close game at Fenway Park that never ends." So here I was with my mother, as close to heaven as I could get.

Dave Morehead, a young arm who'd shown some flash, made his first start for the Red Sox, and the future Hall of Famer Jim Bunning started for the Tigers. The Red Sox were a bad team, the

Tigers a much better team, so the chances for a Sox victory were slim, especially up against Bunning.

But we managed to get one run off of him. However, the Tigers soon tied it up. Arnold Early, a journeyman relief pitcher, came on in the sixth to relieve Morehead, while Bunning just kept on pumping.

By the seventh, Dick Radatz had taken over for Early. Radatz was immense, but so was Bunning. Radatz matched him and did what is now unthinkable for a closer, going seven innings until he pooped out and gave up two runs in the top of the fifteenth. He had performed a Herculean task, shutting out the Tigers for seven innings, but he finally proved to be mortal.

Game over. Totally dejected, I said to my mother, "We might as well leave now and beat the crowd. There's no way they're coming back."

"What are you talking about?" she replied as if I'd spoken treason. "They can still win. The game is never over until it's over."

"Mom, that sounds nice, but they *can't* win this game, and I really don't want to stay and watch them lose."

"Well, I'm staying. You can do whatever you want," she said, and folded her arms in a stubborn pose.

"All right, Mom, but this is painful."

"Just you wait and see. My daddy, your grandfather, Skipper, always taught us that the game is not over until the final out." My mother had a way of giving a person's several relationships when she spoke of them, as in "my daddy, your grandfather," or even "my son, your brother."

"I know, Mom. Skipper used to say that to me as well. But it's just a stupid cliché. I don't care what Skipper said, this game is *over.*"

Still, I couldn't leave without Mom, so, grumpy as hell, I sat watching the Tigers' pitcher take his warm-ups as the infield went through the ritual of fielding ground balls thrown by Norm Cash, the first baseman. They must have felt great, knowing they had the game in the bag. But Mom wanted to stay as a matter of principle. She was nothing if not stubborn.

Frank Malzone, one of my favorite players, led off the bottom of the fifteenth, with a single to left. "You see," my mother said, looking over at me, "there's hope."

"This is *exactly* what the Red Sox do, Mom. They tease you. Just watch. They'll probably load the bases just to lead us on, and then someone will hit into a double play and end it. I refuse to take the bait."

"Just you wait."

The next hitter, shortstop Eddie Bressoud, flied out to left field. "See?" I said. "Can we leave now?"

"Just watch," she replied. On the next play, Red Sox luck (i.e., bad) struck. With Malzone still on first, Bob Tillman lofted a pop-up into the swirling early evening winds of short right field. Dick McAuliffe, the Tigers' second baseman, circled under the ball but a sudden gust took it just out of his reach. However, the great Al Kaline was alertly backing up the play from right field. He easily forced out Malzone, who'd had to hold up thinking the ball would be caught. So now, instead of the Sox having runners on first and second with one out, they only had a runner on first with two outs.

"Can you believe that?" I said. "We get lucky on a pop-up that drops in, but we're up against the best right fielder in baseball who throws him out. That's the Red Sox." Even at my young age, I'd learned how to talk like a long-suffering Red Sox fan.

My mother looked over at me and stuck out her tongue. I actually laughed.

Down to our last out, Billy Gardner tortured us and kept the game alive by singling to left. Runners on first and second with two outs. Chuck Schilling then further tantalized us by driving in Tillman with another single to left, making the score 3–2, Detroit. Now, down by just one run, we had runners on first and second. There was legitimate hope. However, there were two outs, and, much worse, the *pathetic* Román Mejías was due up.

"Oh, God," I said. "Not Mejías. Anyone but Mejías." Over the winter, the Red Sox had traded one of my favorite players, Pistol Pete Runnels, the reigning AL batting champion, for Mejías. Now, just a few weeks into the season, Mejías was hitting .120 and had already proven to be a total bust.

"C'mon, Román," my mother cheered, having alertly picked up his name from the announcer. I was impressed by that. Gotta give her credit. She's into the game. She even got the pronunciation right, so "C'mon" rhymed with "Román."

By now the couple of thousand fans still left in the stands had been suckered into the possibility of a win. These fools, including me, were on their feet, cheering, screaming, yelling.

Román stood in the right-handed batter's box, taking practice swings. *Could it happen this time? Could the pitiful Román possibly deliver?* We all knew the answer was no, but even I was not jaded enough not to root for the possibility. *Maybe this time . . .*

But Detroit had scored against the monster, *Radatz*, the greatest relief pitcher of his era. Once they scored against *him*, fate would deem the game over.

Mejías stepped into the box and looked at strike one. *Of course.* But I was standing up cheering nonetheless. It felt ridiculous, cheering for what the sane part of me knew was flat-out impossible. But that's what fans do. Here I was, only in the seventh grade, already practicing the habit of going nuts, torturing myself by hoping for the impossible. Why did Skipper and my mother think keeping the faith in impossible situations like this was so noble? What was good about keeping faith in a sure loser? What was to be gained in getting your hopes dashed over and over again? Isn't that just stupid?

The second pitch almost hit Mejías in the head. But, damn, wouldn't you know it, as he dropped to get out of the way, the ball happened to hit his bat and glance foul. *Strike two.* "C'mon, Román!" Mom cheered again. She was up, fist raised, yelling as loudly as she could, such a trouper.

Caught up in the collective insanity, I joined in, yelling as loud as I could, "C'mon, Román!"

The Detroit reliever took the ball, rubbed it up, went into his stretch, then reared back and fired what would be the game's final pitch.

Miraculously, Mejías turned on that fastball and shot a rising liner out to left center field, right where the left field wall meets the center field wall. Billy Bruton, the Tigers' speedy center fielder, ran down the ball and quickly fired it back to the infield.

Gardner scored, tying the game, but Billy Herman, the third base coach, gambled and also waved in Schilling. Digging hard, Schilling rounded the bag just as shortstop Coot Veal took Bruton's throw, pivoted, and fired a perfect strike home. But Schilling, gritty as ever, was able to slide home under the tag, barely in time to beat Veal's throw.

Red Sox win! Red Sox win! The team mobbed Mejías. I jumped up and down, hugging Mom. The small crowd left in the park roared as if we were millions. You would have thought we'd won the World Series. As far as I was concerned, we had.

Walking back to the hotel, my mother took obvious pleasure in giving me the lecture I deserved about never giving up, but I was so happy I didn't argue back. She was right. I was wrong. It's never over until it's over.

I presume I'm one of just a few people who remember that game now, more than fifty years later. I told the story of it in my eulogy at my mother's funeral. Not long after, my brother Ben sent me a bat he'd had inscribed with the name Román Mejías.

Miracles do happen. The greatest gift my mother ever gave me, aside from loving me enough to send me off to boarding school and away from Unger and the world of Charleston at the age of ten, was her gift of unshakeable, rock solid optimism. No matter what happened, she never gave up.

27.

My brother John, called Johnny, escaped Uncle Unger and Charleston by going to Exeter for high school and adopting the house on Kettle Drum Lane or Gammy Hallowell's Wianno house as his homes for vacations. He did well at Exeter, winning writing prizes and getting into Harvard.

All was fine and dandy until the week before graduation, when Johnny took the train from Exeter to Boston for a Saturday outing. This was within the rules. However, while in Boston he got drunk—not within the rules. By the time he boarded the train back to Exeter, he was stumbling. His friends loaded him onto the train and tried to keep him hidden in a corner seat.

He might have gotten away with it had not bad luck intervened. It turns out the elderly wife of one of the most senior teachers at Exeter was on that train. Let me call her Eloise Whittington, and her husband D. Loomis Whittington. When Johnny got up to go to the bathroom, he stumbled past Eloise, but she didn't recognize him. She lived off campus and knew few students.

Returning from the bathroom, however, when Johnny drew abreast of Eloise, it didn't matter whether she knew him or he knew her because he suddenly tossed his cookies all over her.

Johnny's friends rushed to intervene and try to make things better but there really wasn't anything they could do other than offer the sputtering Eloise napkins and paper towels, which she angrily accepted. She demanded to know who they were, how old they were, and where they were going. One thing led to another and before you knew it, she had them nailed.

"Wait until Dean Kesler hears about this, young men!" she snapped, heading off to the lavatory to clean up as best she could and try to reconstitute her flagging dignity.

Dean Robert Kesler, with his legendary ardor for expelling students, must have smacked his lips as he listened to Eloise tell him her story. It was just what Kesler needed: an ironclad case for expelling a student mere days before graduation, a student heading to Harvard, no less. That would show all the complacent seniors that they couldn't take graduation—*or the rules*—for granted, no matter what. An ideal lesson for the entire community. No way Saltonstall, the principal and a more forgiving man, could get in the way of this! A perfect kick-off to graduation week.

Although Kesler himself could not expel a student—it required a vote of the full faculty, because Exeter was a faculty-run school—in this case it was a done deal. D. Loomis was revered, and a faculty wife, *any* faculty wife but especially Eloise, wife of D. Loomis, was as protected a species at Exeter as the whooping crane was around the world. It would most certainly be curtains for one John Hallowell.

Enter David D. Coffin, instructor in classics, cousin of William Sloane Coffin, and one of the most brilliant men ever to teach at Exeter. He also happened to be Johnny's adviser. As such, it was his job to advocate for Johnny after Dean Kesler presented the case against him to the full faculty.

Naturally, I wasn't at that faculty meeting, nor was Johnny. I don't know what points David Coffin argued; my guess is he played the troubled-family card in his advisee's defense. But most of all, he must have channeled a pyrotechnic combination of Daniel Webster, Clarence Darrow, and Cicero himself, because, incredibly, against all possible odds, the faculty voted *not* to expel my brother.

How I would love to have seen the look on Kesler's face when the hands went up and the vote was counted! If ever there was a case when Dean Bob, as he was called, could have made book on a conviction ("conviction" was the way he thought), this was it. But David Coffin, and a mercy rarely shown at Exeter, prevailed.

Johnny would go on to Harvard, where he would have a spectacular academic career, as well as come out of the closet. He wrote an autobiographical play called "A Short Safari Through Purgatory," about being seduced by a Harvard Square hustler. A Harvard grad student in English named Ed Hood staged it at Harvard's Loeb Drama Center, and the play received rave reviews.

In addition to developing his life as a gay man in Cambridge, Johnny hyperfocused on his studies. He wanted to excel, and excel he did. Majoring in English, he got such outstanding grades he was elected to Phi Beta Kappa in his junior year, a special group at Harvard called the Junior 16, honoring the top sixteen students in the class.

When I was still at Fessenden, I remember stories of Johnny studying in Gammy Hallowell's attic during most of a January, the so-called reading period before exams. He studied Chaucer so closely that on the notoriously tough final exam given by the quizzical, walleyed Professor B. J. Whiting, Johnny got an unheard-of A+. On the section where you were supposed to identify quotations from *The Canterbury Tales*, Johnny not only knew the source of every quotation, he was also able to supply the line that preceded it and the line that came after.

Harry Levin, who was a world-famous James Joyce scholar and close friends with C. P. Snow, guided Johnny as he wrote his thesis on "Moral Development in the Works of Charles Dickens." By the time he handed in his thesis, not only did he know Dickens backward and forward, he and Harry Levin were banking that he would get the highest grade possible on his thesis. The only obstacle was that Reuben Brower, a notoriously tough grader, would be the main reader of the thesis. I remember the tension building as Johnny asked the whole family to root for him. Finally, in late April, Brower gave his verdict: summa cum laude.

Armed with his stellar record, Johnny went off to London on a Fulbright scholarship where he spent two years studying and, thanks to Harry Levin, all but living with C. P. Snow—Lord Snow—and his wife, the novelist Pamela Hansford Johnson, or Lady Pamela. Johnny would regale me with his animated discussions with Snow and Johnson ranking the greatest novelists of all time. Johnny made the case for Dickens as best he could, but Tolstoy was tough to knock out of first place. The five they settled on were Tolstoy, Dickens, Dostoyevsky, Proust, and Mann. Johnny loved discussions that ranked the greatest whatevers.

Having set his sights on being a star—and becoming one, at least in academia—when he got to Harvard, it was only natural that Johnny would look toward Hollywood when he came home. Although Harry Levin bred him to become an academic, that was just not who Johnny was. In his heart, he loved the bright lights of Broadway and Hollywood, he loved hanging out with stars, he loved fame, and like his favorite author Dickens, he wanted to write for the public, not the academics.

I had been watching Johnny from my vantage points at Fessenden and Exeter all these years. He was enough older than me—seven years—that I never felt intimidated by his academic record, but rather proud.

When he got to Hollywood, he started writing cover stories for *Life* magazine about movie stars. A contemporary of Rex Reed, he got invited to A-list parties and hobnobbed with celebrities such as Bette Davis, Barbra Streisand, Paul Newman, Natalie Wood, Angela Lansbury, Rita Hayworth, Lauren Bacall, Shirley MacLaine, Gore Vidal, and many others. Some of them, especially Lansbury and Vidal, became lifelong friends.

He was riding high. During my first year in college, he published a book called *The Truth Game*, a series of interviews with movie stars. I remember thinking to myself, "His ship has come in." The book was well received, so he continued to work as a journalist and wrote two more books, *Inside Creedence*, about Creedence Clearwater Revival, and *Bodies Beautiful*, a racy novel about a male hustler.

At one point, Jamie—who looked up to Johnny as a role model because, among other reasons, both were gay—went to Los Angeles to visit Johnny in his Malibu apartment. Jamie was hoping to get some guidance from him.

Instead, that visit was a disaster. The worst of Johnny came out as he viciously turned on Jamie, calling him a pathetic loser, a leech, and all kinds of demeaning and undeserved epithets. Devastated, Jamie came back to Cambridge. Amazingly, though, he never stopped caring about Johnny—care Johnny would sorely need in the coming years.

After *Bodies Beautiful*, Johnny's genes, drinking and hard living, and unspecified inner demons caught up with him. He had a psychotic break and was hospitalized in Camarillo State Hospital, near LA.

At one point, Camarillo State had seven thousand patients and was the largest mental hospital in the world. When Johnny was a patient there, it was a highly regarded hospital, albeit a state facility. However, Johnny's memories of it were brutal. He described the electric shock treatments he received as pure torture. This was 1971, and with the enhanced use of anesthesia, it is unlikely the shock treatments were painful, but Johnny loved drama. Also, to be fair, just being in a state mental hospital can be a deeply traumatic experience and can leave permanent scars, as it did for Johnny.

It's not clear to me who took over and brought Johnny home. At the time I was a junior at Harvard, taking pre-med courses and majoring in English, so I was busy, and wouldn't have been the organizer anyway. Even though my father was a former patient in a similar hospital with the same diagnosis as Johnny's, Dad would not have been involved, as he and Johnny were alienated because my dad couldn't accept Johnny's being gay. My mother, God bless her, would have been drinking and not up to the task, so I am pretty sure it was Duckie who took over. When I was growing up, she was the one who was always there when help was desperately needed.

Having before known only high-achiever Johnny, I was unprepared for how crazy he was. Once the arrangements were made, I went to Logan Airport with Duckie and Jamie to meet him. Johnny was so out of it I was surprised he had been able to make the flight.

In those days, I was a smoker. When we met Johnny, he saw a matchbook I had that said "King Edward" on the cover, an ad for King Edward cigars. Rather than say hello, out of nowhere Johnny snapped at me, "Oh! So you're now the *King*? *Edward*?"

At that moment Johnny provided me with my first up-close experience of a psychotically paranoid person. Duckie calmed him down but I was totally taken aback. His eyes were flashing, he was looking all around, he was talking ragtime-speed about spies we should watch out for, and he was targeting me as his enemy.

This was not the brother I knew—not the Johnny from Camp Kabeyun, not the Johnny who wanted to take me to a Red Sox game, not the Johnny who'd debate with me the primacy of image or word in modern culture, not the Johnny who helped support Mom when he was making a good living in Hollywood.

I almost thought he was putting on an act. Maybe I hoped he was putting on an act. But no one could fake what he was doing. He was like a dog gone rabid. In years to come I'd discover how standard this kind of reaction can be among psychotic people, but right then it was anything but routine. I'd also later discover not to fear it, that it can be managed. But I knew none of that at the time. It was horrifying to see my brother change from someone I loved and who loved me back to someone who hated me and whom I simply did not know.

I marveled at how Duckie kept her cool, gently coaxing Johnny to sit down in one of the chairs in the airport waiting area. I took the hint and backed off while she and Jamie talked Johnny down, at least enough for him to ride in the car.

Johnny was admitted that day to the psych unit at Cambridge City Hospital (now Cambridge Hospital). He was started on anti-psychotic medications, which later gave way to lithium, the same medication that saved my father. His first psychiatrist was a brilliant doctor named Steve Stelovich, who actually did a few sessions of family therapy with Johnny, Jamie, Duckie, Uncle Jimmy, my mother, and me. I have virtually no memory of those sessions, other than a visual of the seven of us sitting in a circle in folding chairs in a conference room on the ward. We only met once, as I recall, since it was a minor miracle that Uncle Jimmy had left Chatham to come up to Cambridge even that one time.

I wish we'd met more often. There was so much for us to talk about. But it was not to be. While as a family, we were great talkers, we were not disposed to sit down regularly with a trained professional and talk things out. Few families are.

So began Johnny's long life as a chronic mental outpatient. He would never regain the intellectual, creative brilliance he once had, but only rarely would he display the extreme nastiness and selfishness he'd flashed during Jamie's LA visit.

He was not as lucky as my father, who regained full function, although not as at high a level as his brainpower would have warranted before he had his psychotic break. Johnny never got back his mental fastball, or even a reasonable facsimile.

Whatever the cause, Johnny's safari into madness left him permanently changed. He was diagnosed with bipolar disorder, as well as alcoholism.

As an outpatient under the care of his new doctor, Dr. Fields, Johnny joined AA, became sober, and remained sober the rest of his life. He was as devoted a member of the AA group as he had been a student of Chaucer. He learned the lines backward and forward. One day at a time. Hurt people hurt people. Live and let live. Easy does it. Poor me, poor me, pour me another drink. Count your blessings. He could recite scores more. I actually loved hearing them.

While in AA, Johnny met a woman who would become his lifelong companion. Ann was a few years older, was divorced, and had two grown children. She was thoroughly heterosexual, Johnny was thoroughly homosexual, but the two developed a bond stronger than most marriages. Although they never had sex with each other, or anything close to it, they loved each other and lived with each other in an apartment in Cambridge for decades, until they lived together in a nursing home.

It gave me the creeps to be around Ann because she was so theatrical, always insisting on being the center of attention. When I would visit them, she'd usually be lying on her bed in her nightgown. She had big hair, which Johnny helped her put up every day, and lips covered in bright red lipstick. She always insisted on a kiss. It was all I could do to bend down, put my hands on her shoulders, and kiss her cheek.

Johnny turned Ann into his star. He was utterly devoted to her. He waited on her hand and foot, literally, rubbing her feet whenever she wanted, and putting whatever she asked for into her hands. He all but fed and toileted her, and would have done that had she bidden him to. He would roll his eyes, but we could tell he loved being at her beck and call. She was a caricature of my mother.

Loving literature and the movies as much as Johnny did, Ann would ask John, as she called him, to read to her for hours on end, or rent movies. Since Johnny was a living encyclopedia of all things Hollywood, he could regale Ann with endless stories about the movie stars he knew, as well as tell her the abundant gossip he'd picked up during those years. "Paul Newman really was as good a guy as he seemed"; "Bette Davis was the same off-screen as on, and you damn well *better* never upstage her"; "Loretta Young had a swear box on her set. If you said 'damn' or 'hell,' you had to put a quarter in the box. One day Ethel Merman was on set and she put her hand on her hip and shouted over to Loretta, 'How much will it cost me to say *Fuck you!*' "; "The day I backed my car over Natalie Wood's cat and killed it, I thought she'd never talk to me again, but she came around"; "The best, kindest, most genuine person in all of Hollywood or acting anywhere is Angela Lansbury"; "People would try to pit Rex Reed and me against each other, but Rex was actually really nice to me"; "You wouldn't believe how many famous actors started off as male hustlers"; "I don't know why some people don't like Barbra Streisand, I love her"; "Ella Fitzgerald was the most talented impossible interview in the world; she couldn't put a sentence together"; and on and on. There wasn't a star he didn't know something about, and he could recite all the Oscar winners going back to the beginning. Had he not gone crazy, who knows what he would have done out there?

But he was happier, it seemed to all of us, living with Ann than he had been in Hollywood. He'd go to AA meetings, go to see Dr. Fields regularly, take his lithium faithfully, and never touch a drop of alcohol. He did not relapse once.

As the years went on, conversations with him became predictable to the point of being tedious. For example, no matter how many times I, or anyone else, would tell him that we liked Bette Davis, he would ask, in almost every conversation, "Do you like Bette Davis?" or "Do you like Barbra Streisand? I do." Until the day he died, he asked virtually every visitor, "Do you like Bette Davis?"

It's ironic, of course, because none of them cared a hoot about him. After he had his crack-up and left Hollywood, none of these stars he so cherished checked in on him—with one exception.

Up until his death, Angela Lansbury—Miss Lansbury, as Johnny called her—kept in touch, sending him notes, inviting him backstage when he'd go to New York to see her on Broadway, always asking after Ann's as well as his own health. No year passed, rarely even six months, without Johnny's hearing from Miss Lansbury. Her attention meant the world to him.

Other than alcoholism, I don't know what Ann's diagnosis was. Whatever it was, it fit perfectly with Johnny's. When he came to family events, he'd have to call in and check on Ann back at their apartment every hour or two. If not, she'd call wherever he was, annoyed. Rather than get angry back, Johnny would lovingly reassure her that he'd be home before too long.

One of the problems of people who have a chronic mental illness is often poor personal hygiene. Both of them struggled with this issue. Thankfully, there were visiting caretakers who'd come bathe them as they got older.

I tried to keep my relationship with Johnny as normal as I could. I tried not to treat him as a patient but as a brother. So when he'd ask me if I liked Bette Davis, I'd reply, "Johnny, you know I do. You've asked me that question a thousand times, if not more. Can't we just stipulate for the jury that I like Bette Davis so you don't have to ask me again?"

He'd laugh. But it didn't work. The same questions always came.

Sometimes I'd try to tap into his former brilliance. "Why was Dickens so great?"

"Characters," Johnny said. "Except for Shakespeare, he created the greatest gallery of characters in all of literature. And his moral sense. He always took the side of the underdog. He had a profound sense of injustice."

It was still there, although fading daily. You just had to look for it. For a number of years, after he was stabilized with the right medications, he taught writing at the Harvard Extension School. He was a beloved teacher, mainly because he regaled his students with stories of Hollywood stars.

He also knew how to teach writing. From his days at Exeter, with great teachers like George Bennett, Henry Bragdon, and Colin Irving, and then tough editors at *Life* magazine, Johnny knew the craft of writing well. He asked his students to write regularly, carefully going through their papers and always offering encouragement.

His decline happened over years, but Johnny's slow slide from his peak stirred in me both pity and fear.

It was beyond sad—at least for us who loved him, I'm not sure he, himself, cared all that much—when both his physical and mental health failed to the point that he could no longer teach. Undaunted, he made his rounds from the apartment he shared with Ann on Walden Street in Cambridge to the neighborhood coffee shops and corner stores.

Having Parkinson's disease, being mildly incontinent, being seriously overweight, and never having been terribly agile to begin with, his gait was halting. He took small steps (always had—the family made fun of him for this even when he was a teenager) and had to stop often and regroup. But unlike Ann, who had to be carried or driven everywhere, Johnny would walk. No matter how long it took, he made his rounds.

His favorite haunt was a coffee shop called Simon's next to a video rental store called Hollywood Express. The staff there knew him well. He came in every day. He'd tell them all about the movie stars he knew. The people who worked there ate it up, until, after a couple of years, it got old, and Johnny's personal hygiene declined to the point where he emitted a foul odor. Jamie, who sometimes accompanied Johnny, would get calls from the store asking him to do something about it.

The tables were now completely turned from the days in California when Johnny was insulting and demeaning Jamie. Now Johnny needed Jamie. Jamie, ever one to put himself down, cynically said the only reason he was helping Johnny was to take pleasure in being in the power position. But I know Jamie better than that. He was doing it mostly out of love and sympathy for Johnny. Being gay and coming out long before it was fashionable, Johnny had paved the way for Jamie. Of course, he had been cruel to him as well. But Jamie hung in there no matter what.

Lyn and Tom were far enough away down in Rehoboth, just outside Providence, that they were not on call for Johnny, although Lyn took a strong interest in him, so much so that she organized a thorough cleaning of Johnny and Ann's apartment. True to form, she swung into action and got Tom, Jamie, and me to meet at the apartment and have at it.

This project was more an excavation than a cleaning. Literally years of dust and grime had to be all but peeled off the inside of lamp fixtures, the tops of bookcases, the areas under pieces of furniture. The inside of the refrigerator resembled a compost heap. The bathroom was not all that bad because of the visiting nurse who came regularly to bathe Johnny and Ann. But the rest of the apartment needed fumigation.

It took us all day, but, some twenty-five garbage bags, endless rolls of paper towels, and countless sponges, solvents, detergents, brooms, vacuumings, and gags later, the place looked and smelled pretty good.

Ann lay regally in bed during the entire process, offering suggestions from her throne, while Johnny scampered around trying to help but mainly just getting in the way.

Lyn was tough with him. "Go sit there," she would command. "No, I don't give a flying fuck about Barbra Streisand." None of this bothered Johnny. No matter how hard Lyn came down on him, he kept on smiling. I believe he regarded her snapping at him as what a diva would do. He was used to divas.

When the cleaning was finally finished, Johnny walked us to the door. Ann requested that we all come in and kiss her goodbye, which Jamie and I did, while Tom and Lyn made a beeline outside with the trash before Ann could stop them.

Worn out, we went to a nearby bar and talked about how it can be that two people could live in such squalor, how was it possible for Johnny to love Ann as much as he did, how glad we were that the two of them had each other, and that Halley's Comet would come before we did a clean-up job like that again.

Still, the question remained, what could be done? One cleaning wouldn't do the trick. Lectures didn't work. Being told by Simon's to stay away only worked for a little while until Johnny went back. As a family, we were learning firsthand what it's like to care for someone with a serious, chronic mental illness.

We had to remind ourselves of that fact. All of us would forget that Johnny was seriously disabled, for reasons beyond his control, and we'd blame him for being a slob, for not taking care of himself, for being a disgusting mess. It was hard for us, sometimes even for me who was in the business of knowing better, not to chastise him for being impaired, and for getting worse. Instead of giving him the credit he deserved for consistently staying sober, I often forgot and laid blame on him.

The one who didn't was the woman I would marry, Sue, who loved Johnny as if he were her own brother and took care of him and Ann to the point of taking them shopping for clothing, making sure they got the right medical care, looking after their finances— in short, along with Jamie, doing everything I should have done but didn't. That's just how Sue was and is with everyone.

Johnny died on Christmas Day 2014, in the nursing home he and Ann ended up in, at the age of seventy-two. I think he chose that date to honor his great hero, Charles Dickens, and his favorite story, "A Christmas Carol."

28.

After the divorce, my mother lived in a series of houses in Chatham before Aunt Janet found her a placement in publicly assisted housing. During school vacations I'd stay with her, or with Duckie and Uncle Jim.

I looked forward to coming home to Chatham, mainly because I'd see Jamie and Lyn. Mom, as much as I loved her, was not so easy. She wanted me to pay attention to her, but by early evening she usually had had so much to drink that it wasn't fun to talk with her. She'd launch into a long series of complaints about her life: how Duckie was too controlling, Lyn influenced me too much, I wasn't nice enough to her. It was not easy to listen to.

I could tell when the switch flipped, when she was no longer remotely sober. I tried to cover up for her with Lyn and Jamie, telling them she was just tired, that she hadn't had all that much to drink, that she was doing her best. But I knew what was going on. I wanted to protect her, but even more, I wished she'd quit drinking.

She would tell me, "I can quit any time I want to."

"Then why don't you? We all wish you would."

"Who do you mean, 'we all'? If you want me to quit drinking, I am happy to."

And then she'd go on the wagon, sometimes for as long as a month, but usually just for a few days. Those chunks of time were great. She was fun to be with, she could make us all laugh with her stories, she was the mother I loved.

The house where she lived for the longest time was on Old Harbor Road. Her room was upstairs, mine was a corner room on the first floor. Next to it was a sitting area, which led to a small kitchen and a larger living room.

During vacations from Fessenden and Exeter I developed my idiosyncratic routines. When I wasn't at Jamie and Lyn's house, I'd make myself a minute steak at Mom's little house. When I worked at Bearse's grocery store I learned about minute steaks. One of my responsibilities was to take slabs of chuck steak and feed them through pronged rollers that cross-hatched the steaks to be so thin and tender you could cook them in a minute.

I would bring a few home with me, along with some canned asparagus and store cheese, which I'd slice off the big wheel of cheddar kept atop the meat counter. My favorite meal became a couple of minute steaks, a piece of toast, canned asparagus with a spoonful of mayo on top, and a chunk of cheddar. I had a friend home from Exeter one vacation and offered to make him this meal, but he said it sounded disgusting. For whatever reason, I loved it, and I ate it whenever I was alone. It's one of many idiosyncrasies I developed growing up that most people find weird or worse.

Christmases we'd have at Duckie and Uncle Jim's house. Jamie, Lyn, and I would spend the entire school vacation leading up to the holiday shopping, decorating, getting into the spirit. Lyn was the cheerleader, and even though she was pretty controlling, she really did make it fun.

Mom would drink too much to go shopping. I don't ever remember her Christmas shopping. She'd ask Duckie to buy us presents, and one year she forgot altogether.

One Christmas Eve, I was at Jamie and Lyn's house and she phoned me. She'd had too much to drink and slurred into the phone,

"Can you go out and buy some Christmas presents for me? I will pay you back."

"Mom, it's ten o'clock at night. The stores are closed."

"No, you don't unnerstand, I don't have any presents to give to anyone. Can you go buy some for me, please?"

Jamie and Lyn, and Duckie too, felt angry, and rightfully so, but I couldn't get angry. Mom just seemed like such a sad figure to me, and I had to stick up for her, but that Christmas Eve night might have been the last time that I let myself feel that much hurt for her. After that, I kept loving her but sealed off the bleeding.

29.

I arrived at Exeter in September 1964, just a few months before my fifteenth birthday. My brother John had graduated a couple of years before I arrived, and our grandfather, Skipper, had gone to Exeter as well. But it was not the school Jamie attended; the Hallowell side of the family preferred Milton Academy because Gammy Hallowell's father had taught Latin there.

Having been first in my class every year at Fessenden, I was unprepared for the quantum leap in smarts that I would find in the kids at Exeter. Suddenly I was not number one, far from it. I had to work hard to get B's, never mind A's. While it was jarring not to be number one, it didn't upset me too much because we were all in the same boat: shocked to discover that no matter how smart you might think you are, there are many people who are smarter.

Exeter cast a spell over me, over most of us. We entered as unsophisticated ninth graders, called "preps," believing that we were smart because that's what we'd been told and we'd demonstrated at our previous schools, having no clue, really, how to think or write and certainly not the remotest idea of how deeply the Phillips Exeter Academy would change us.

It broke some kids. Exeter could be a cruel place. Many teachers delighted in giving D's or F's, as if they were upholding the academic standards of the world and preserving Western civilization. We also had Dean Kesler who delighted in expelling kids. He once gave a chapel talk on the felicitous effect of being expelled from Exeter. (Each day started with a required all-school meeting called "chapel," which began with a hymn but was in no other way religious. It included a talk by a faculty member or some guest.) Dean Kesler kept a polished rock as a trophy on his desk, given to him by a student who'd been expelled only to come back years later and tell Kesler that being kicked out of Exeter had been the key to his success. During my class's four years, 20 percent of my classmates were expelled and another 5 percent left because they just couldn't take it.

Exeter was in flux during those years, as was the rest of the world. We were plunging into the heart of the 1960s. Exeter's new principal, Dick Day, started right along with us. He would bring coeducation to Exeter, along with a new library, the end of required church, and a complete review of the curriculum.

But I didn't care about any of that. I was concerned with the meaning of life, trying to figure out what to do with mounting sexual feelings in the absence of girls, how to get good enough grades to get into Harvard (the preferred destination for almost all of us), how to speak up at the round tables that provided the setting of all of our classes, and how to stay happy in a place that seemed quite oblivious to the happiness of its students.

It was a sink-or-swim place. Our dorm head, Ted Seabrooke, legendary wrestling coach and a hero of John Irving's, gave us monthly pep talks at dorm meetings. "Exeter gives you just enough rope to hang yourself," he'd tell us, then chuckle, "and there's a lot to recommend that. I'm not kidding, though. No one is going to come around and make sure your homework is done. You can waste *all* your time if you want to. Nobody's gonna tell you to go study. But you won't last long here if you don't work. Just don't think we'll keep you afloat if you start to sink, because we won't."

I'd listen to him and think to myself, *Well, I will do my work, so I won't sink.* Other kids would get scared and shut down. Not many, but a few. Seabrooke also cautioned us about our record: "Your record follows you everywhere you go. Life is starting to get serious now. It's time for you to get serious, too."

As severe as all that sounds, the fact is Ted Seabrooke had a heart as big as the world. He was a fantastic dorm head. Rather than frighten me, his monthly pep talks inspired me. Just as he coached his wrestlers on how to pin rather than get pinned, he coached us on how to do Exeter right, and by extension how to do life right.

Those talks provided stability, a feeling of confidence: I could make it as long as I kept working hard. I knew talent alone could not carry me. What his talks could not reach, however, were the dark moods swirling around inside of me.

They'd started at Fessenden, those nights when I'd stay awake praying Mom could be happy. But when I hit Exeter, the bleaker side of me took serious hold. Exeter did that to many of us.

Not everyone felt this, of course. The student body was made up of so-called "posos" and "negos." The posos were the positive ones, the jocks, the optimists, the student council types, the cheer-leaders, the ones who bought the motto of the school, inscribed in marble over the entrance to the Academy Building, where we went every morning for chapel: HUC VENITE PUERI UT VIRI SITIS—"Boys, come here that you may be men."

The negos were the cynics, the artsy types, the sophisticated kids, the ones who had mastered the art of being cool, disdainful, and blasé. They snickered at the school motto, looked critically at the privileged base most of us came from, smoked cigarettes and soon marijuana, and in general mocked pretty much everything most people admired.

I was part poso, part nego. By the time I graduated I was pretty firmly poso, but I thoroughly understood the position of the nego. Exeter forced many students into that mold, if only to defend themselves from the daily assault on self-esteem the school could mount.

The sadness I felt at Exeter was the culmination of my first fifteen years of life. The chaos and hard times that had surrounded me in Charleston left me insecure, uncertain about what to expect next.

Furthermore, happiness and mental health simply were not WASP desiderata. Character is what counted. My aunt Nell, for example, as learned and well-read a woman as I ever met, would have scoffed at the concept of self-esteem. She emphasized humility, restraint, discipline, deference, moderation, abstinence, all subsumed under the heading of character. When we took baths under her supervision, we were allowed no more than a half inch of water in the tub. It was against her better judgment that she allowed the water warm rather than ice cold. You could say some propensity toward a gloomy disposition was bred in our bone.

I remember going alone to the Lamont Art Gallery on a gray and snowy Sunday in February at Exeter to watch Ingmar Bergman's movie *Wild Strawberries*. It was in Swedish, black and white with English subtitles. It was about an old man, a doctor, revisiting his past in various ways.

The gloom of the movie totally captivated me. *Yes, yes*, I felt, *that's life*, even though I had little idea what was going on in the film. It affected me deeply, immersing me in gloom. That movie provided an objective correlative—to use T. S. Eliot's cumbersome term, which I use here only because I learned the term that year at Exeter— for the darkest, saddest parts of my being.

Oddly enough, the final scene in the movie depicts the old man going to sleep in his bed, a smile on his face. Somehow, after the ordeal of his journey, he finds peace at the end. Now, some fifty years later, I wonder if that final scene didn't work its way into my unconscious, planting a seed of hope I did not feel at the time.

After the movie, I walked alone back to my dorm down the blacktopped paths of Exeter under quietly falling snow, flakes dropping aimlessly, drawing me into a dream state akin to what the movie had induced. Crossing Front Street beneath the flashing yellow light that hung over the crosswalk in the gray winter twilight—I could hear the faint click the light made each time it flashed—then stopping for a moment to button up my jacket as it was getting colder out, I realized I felt as alone as I'd ever felt in my life. I felt like that old man in the movie, walking through scenes I didn't understand.

I sometimes revisit that walk back to the dorm after viewing *Wild Strawberries* and think of it as an enactment of a part of myself that got born at Exeter, that the movie midwifed, the empty, confused, lost, depressed, despairing, hopeless part of myself that I have spent so much of my life fleeing, denying, repudiating, or in some other way trying to destroy or, at best, to minimize.

And yet, in that young man taking his lonely walk in the near-darkness of New Hampshire's winter, I see a stronger part of myself slowly growing, learning to bear the harder parts of life without sugar-coating, the parts that I'd tried, in keeping with my mother's optimism, to pretend didn't exist. I see a part of myself that could look up at the night sky and see absolutely nothing at all, no meaning, no reassurance, nothing but distant, indifferent stars, and not recoil in terror, but just see it as it is, like the click of the flashing yellow light.

The walk took only ten minutes or so. Soon I got back to the busyness and bright lights of my dorm, Bancroft Hall, which at that point felt like a banquet hall beaming light and life in a dark surround, with the sound of *Star Trek* on the TV in the common room, the clang of dishes coming from the kitchen as dinner was being whipped together and served up, the voices of my friends talking in the common room—in other words, the sounds of life, of purpose, of diversion, everything I needed after *Wild Strawberries*.

This was Exeter: meaning and meaninglessness; poso and nego; my dream to set the world on fire and my fear I could never do it.

30.

It wasn't clear to me what Mom did during the day. It seemed against the laws of nature that she would be employed. Duckie worked as a real estate agent, driving all over town with the help of Jamie and me to open and close rentals during the summer; Janet worked at the town offices; but for some reason my mother never worked. We never even challenged her on this.

The more time passed, the more she became wrapped up in her daily routines and drinking. She pretty much lost interest in what I was doing, although she would ask. She just wouldn't remember how I replied.

One night my freshman year at Harvard, she phoned and asked me where I was. I told her I was in my dorm room. "Oh, at college?"

"Yes."

"Where do you go to college?"

"I go to Harvard, Mom, just like Jamie and Johnny."

"Oh, that's good," she said. "Why don't you call me more often?"

"I will. Why don't you turn off your light and go to sleep now?"

In late afternoon, Duckie would come over to Mom's house and the two of them would drink their bourbon on the rocks with water and talk, usually quarrel a bit, but also gossip. My mother would often complain to Duckie about me, especially when I got to college and became more distant, how I wasn't spending enough time with her.

I didn't have any friends in Chatham except Jamie and Lyn, so there was nothing for me to do except go to their house or watch

TV. When I got my driver's license, I'd drive my mother's car to their house every night I could, and come back late, by the time Mom had gone to bed.

One night—I was still at Exeter, so I guess I was sixteen—I came in around midnight and heard noise from Mom's room upstairs. I went to the foot of the stairs and called, "Are you awake?" When I heard more noise, I went up the stairs, steep Cape stairs with about twelve-inch risers. When I reached her bedroom, I saw my mother roll over in bed toward me, bare-shouldered, and say with a slurred voice, "Neddy? Is that you? Are you home already?"

A man next to her got out of bed and started putting on his white Jockey underpants. I saw his pasty naked butt in the moonlight.

"Steve was just giving me a back rub."

I turned around, went back downstairs, and took a seat in the living room. I'd never seen my mother in bed with anyone except Uncle Unger. Seeing her that night shook me. I remember saying to myself, *She has every right to sleep with whoever she wants to, why should you be upset by it, she's lonely and it's a good thing for her to have a man over.*

Still, I felt upset. When Steve, who turned out to be her hairdresser, came downstairs, he handed me a glass of whiskey. "Here, drink this."

"Thanks."

"Would you like to talk?"

"No, thanks."

"OK, well, I guess I'll be going then."

"OK." I drank the whiskey and sat for a while staring at the wall, letting the alcohol work its magic. Then I called Lyn. She came over and we went for a drive.

Lyn and I drove around for a while. I don't remember a word she said, just the feeling I had of immense gratitude that she was there. She looked out for me so many times, that was just one of the more poignant for me.

The next morning, Mom and I didn't talk about the visitor from the night before. I'm not sure she remembered. I hoped she didn't.

31.

In April 1966, *Time* magazine, which back then was the magazine of record, read by everyone, published its most famous cover ever, a black background upon which was written in red block letters, one word to a line:

IS

GOD

DEAD?

Had they been polled, many if not most of my classmates would likely have said yes, on most days myself included. But that made me feel like a traitor. God had had my back in Charleston, so why was I turning my back on Him now? Why was I starting to side with Aunt Marnie and believe that after death there is only blackness?

Perhaps with this issue in mind, Exeter required students to attend church. Most of us didn't want to go, preferring to sleep in on Sundays. Many resourceful students discovered a way out: If you declared yourself to be Jewish, you could go to a Friday evening service, fulfill your religious requirement, and sleep in Sunday mornings. As word of this loophole got around, the Jewish population at Exeter skyrocketed, with the shorter Friday evening service supplanting church for many students.

But I opted to get up Sunday mornings and go to Phillips Church—in part because of my emotional ties going back to St. Michael's in Charleston, and in part because of the school minister, Fred Buechner. I loved his sermons. After Exeter, Fred would go on to have an illustrious career as a preacher, a theologian, and a novelist. Widely regarded today as one of the most inspired and illuminating voices in religion, Mr. Buechner, as I knew him then, was special to me only because he held me unblinking on Sunday mornings with his earthy, passionate interpretations of the Bible and of life.

He preached more like a poet, or a fellow sufferer of this state called life, than as a man up in a pulpit delivering wise words. His words were wise, for sure, but his wisdom felt earned to me, as if he'd grappled deeply enough that we really should listen to what he had to say. We were lucky to have him up there, but we didn't know how lucky then. That's the way a lot of Exeter was; some improbable twist of chance provided us with far more than we ever knew we were getting.

Coming under the influence of Fred Buechner while I was gestating my faith, while my hopes were daily tested by the smart and merciless cynicism that burned through Exeter, listening, rapt, to this Lincoln-like man who had hard-won knowledge of whatever it is that lies beyond knowledge, grabbing on to the sermons he cast out to us like phosphorescent lifelines in the gloomy zeitgeist of those years, watching this man after each service embrace his wife and hug his three daughters as if they were his everything, well, to tell you the truth, it made me believe in God.

This is what set me and church apart from most people I knew, starting with St. Michael's: I felt good there. I looked forward to seeing Mr. Buechner in that pulpit, and getting giddy on his words. There, I felt happy. None of the fear and guilt so many people associate with church and religion. *Exactly the opposite.* I found church a joyful place, full of hope and love and the grandest, most transformative feelings and ideas. You knew, or at least I knew, what you were going to get in church, but you would also be surprised, in good ways. It had very little to do with belief. I could far more easily have argued against the existence of God back then, or now, than in favor. And yet I loved church. I loved the sanctuary. It was the one place where what matters most in life stood front and center.

Fred Buechner must have felt a calling as school minister preaching to a smart, largely agnostic if not atheistic student body to make a case for God and faith, to give us maybe our last, if not only, chance to take religion seriously, to be surprised by God.

But I came to church for something different, even though I didn't know it at the time. I thought I was going because I loved Mr. Buechner's sermons. But actually, I was there to get dosed up on feelings I didn't often find at Exeter. Even though I do not believe giving love was Fred Buechner's primary objective, that's what he gave me, whether he meant to or not. Love and hope.

Of course, Mr. Buechner was very smart. At Exeter you had to be smart if you wanted to be taken seriously. To make the case for God in 1966 at Exeter, you had to be one clever person indeed. But smart didn't matter to me. What mattered to me is what came through to me from Fred, week after week, his absolute conviction, his bone-honest testimony, that life was good no matter what.

Because he'd written about it, I knew that Mr. Buechner's father had committed suicide when Fred was ten years old. That he could be so hopeful, so bold, so loving, so sure of the goodness in life in spite of the doubts and fears that that suicide must have instilled—that made me believe that I, too, could make my life good.

Without knowing me at all, without ever having a single one-on-one conversation with me, Frederick Buechner changed me that year, 1966, the year of the *Time* cover, marking me forever. Then summer came, and Mr. Buechner was gone. He moved on from Exeter to devote more time to his family and his writing. This happened two years before Exeter's trustees, in keeping with the times, abolished the requirement of students attending any religious service whatsoever.

32.

The summer of 1966 would see my life, and the lives of my entire extended family, change dramatically. Gammy Hallowell died.

I was in an upstairs bedroom in the house on Old Harbor Road when, in the middle of an August afternoon, Mom called up to me, "Gammy Hallowell just died."

I called back down, "OK." That's all I had to say. But it was so not OK. I lay on my bed and stared at the white ceiling lit by the afternoon sun coming in through the windows. *What happens now?* I wondered. I felt my world go flat, as if going from color to black-and-white.

Gammy had not only supplied the wealth and opulence, she'd been the matriarch. She ruled. She made things special. Truly a grande dame, she'd married James Mott Hallowell, a man nineteen years her senior, and had their three children, James, Nancy, and Benjamin, only to see her husband die when he was sixty-three and she was forty-four.

Undaunted, she picked up and charged ahead, running the garden club, being a regular at the country club in Brookline, hosting duplicate bridge tournaments at her home, supporting the Faulkner Hospital and hosting an annual picnic for all the student nurses at her palatial home on the water in Wianno, on Cape Cod, active (her operative adjective was "active") in all sorts of projects, causes, initiatives, rummage sales, and elections.

For a number of years she carried on a romantic relationship with Charles Stetson, a distinguished Boston attorney who was big in the railroad business. He'd take a two-car train to the Cape from Boston to visit Gammy, his Louise. One car was for him, the other car for his hunting dogs. He loved to hunt black duck, which my dad loved to do as well. Gammy used to be famous for roasting the ducks he'd kill; she'd roast them in the oven for just one minute, so the blood would still spurt out of them.

Having lost first her husband and then years later Mr. Stetson, Gammy made it the rest of the way without a significant man in her life. Her eight grandchildren became her focus, evidenced by the bracelet she wore with each of our names on it.

Trips to Gammy's house for Christmas and weekends in the summer were extra-special. She was wealthy enough to afford a huge house, beautiful grounds, a croquet court, a rose garden, a sailboat moored at her own dock, vegetable gardens, two cooks, a gardener named Manny who ate raw garlic and had a long gray beard, Luther, the chauffeur, and various part-time help for an assortment of chores.

I was always allowed to order my breakfast as if I were in a restaurant. My favorite was scrambled eggs, because Girda, the Swedish cook, would make them not with milk but with cream.

After dinner, we'd have finger bowls. The first time one of those was put down in front of me, I nudged Jamie, who was sitting next to me, and asked him if this were all we were having for dessert. "No," Jamie said. "Just watch what Gammy does and you do the same." As always, he looked out for me.

She paid all our tuitions to the schools we attended, and left money to cover all future education, which in my case included medical school.

After Gammy died, she'd provided for us financially, but what she couldn't provide was someone else to be the matriarch, the convener, the arbiter of taste, the great lady that she'd been.

I have a hunch Duckie and Nancy's husband, Dick, were not all that upset to say goodbye to her, as she could be a fierce mother-in-law, but she remained loyal, no matter what.

I pretty much lost touch with Nancy and her family after Gammy died because there was no longer anyone to gather us together. I'd no longer visit that magical house on the point in Wianno, where I'd see my dad, play croquet, learn how to fish and sail, play canasta, and eat scrambled eggs with cream.

Decades later I went back to see the house. I wish I hadn't. I should have known better. Rich people who buy and remodel old family homes have a special knack for surgically removing all the character, charm, beauty, and spirit the old place once had, replacing that with garish, grotesque displays. The people who bought Gammy's estate butchered the place, of course. I am only glad she was not alive to see it.

33.

The summer of 1967 was one of the most magical periods of my life. Three unpredictable things happened that brought me great joy.

It was the summer between my eleventh and twelfth grade years at Exeter. As part of a program called Crossroads Africa, I traveled to Togo, on the coast of West Africa, to build a schoolhouse in a little seaside town. The program was similar to the Peace Corps, only it lasted just the summer. Five college students, plus me, a rising high school senior, joined our leader, Father Frank Sullivan, the Catholic chaplain from Lehigh, to fly from Washington to Accra, the capital of Ghana, then to Lomé, the capital of the French-speaking Togo, and then get into a Land Rover to drive to our encampment on a modest military base next to the tiny village of Glidji.

The first surprise was falling in love. Well, sort of. I'd been in love with love but had never kissed a girl, let alone made out or had sex. Sadly, as much as I wanted to, I didn't do either of those things either that summer, but I did develop feelings for a woman which I had never felt before, so I called that falling in love. I so wanted to fall in love, well, why not call what I felt for this beautiful, smart, talented woman love? If it wasn't true love—it was unrequited and not physical so I guess it can't really be called true love—then it most definitely and assuredly was a deeply passionate crush.

One of the five college students in our Crossroads Africa group was Susan. She was twenty-one. From day one, she captivated me. She had a full and open face, beautiful large brown eyes, and a smile that seemed to gather up and radiate all the warmth in the world.

Not since age eight, when I'd had a crush on Tinka Perry, had I felt anything remotely similar.

OK, so I was seventeen, and what could the romantic stirrings of a seventeen-year-old be other than lust? But then, Romeo wasn't much older than seventeen, was he? Well, if this wasn't true love, it was true lust for sure, as I was totally turned on by Susan. She was *zaftig*, to use the Yiddish term, not fat but curvaceous and welcoming. She played the guitar, had a beautiful singing voice, was deeply spiritual and liked to dance. I didn't know how to dance, but for Susan I would try. She liked to talk about theology and religion, which I did as well. She liked to talk about life and politics and art. It didn't matter to me what she talked about, I was ready to listen and join in.

I would not have let my being seventeen and her being twenty-one hold me back, though, were it not for one other detail. She was engaged to be married. How I hated that guy. I pumped her about him, hoping to find or instill mixed feelings or some crack in the relationship. There was none and I was unable to create one. Susan would talk on and on about how much she loved this man, and I would listen. What a dope.

I talked with Frank Sullivan about it. Frank was used to counseling young people, as this was part of his job as Catholic chaplain at Lehigh.

"Have you been in love before?" he asked.

"No. I really have almost no experience with girls."

"You better back off," he said. "You'll only end up getting hurt, and you will disrupt the group. You know we're here to build a schoolhouse, not to practice falling in love."

Duly warned, I did back off as he advised, but I still rode the wings of love for the rest of the summer. I would think of Susan on my way to sleep every night.

Unless I was thinking about the second surprise of that summer. Like my love for Susan, my love for the Red Sox was also frustrated love because the team never won. All of Boston lived by the cry "Wait 'til next year!"

While I was in Togo, the team changed the face of Boston baseball forever. The summer of 1967 came to be called the Impossible Dream season, ushering in the decades of contending teams and sold-out games Fenway Park has boasted ever since. The Red Sox changed from a sleepy team no one wanted to see, except old geezers and little kids, to the hottest ticket in New England.

And I had to be across the world while it happened! Each day in Togo I'd get reports and shake my head, not knowing whether to kick the dirt or jump for joy. Is this for real? Like a true fan, I assumed it was my leaving the country and traveling to Africa that had triggered the Red Sox's success. Maybe I should live in Togo for a few months more. Wouldn't my returning home jinx them?

That's how sports fans think. Each of us, deep in our deranged brain, harbors superstitions we turn ourselves inside out trying to honor. A true sports fan has a socially acceptable mental illness, a chronic, unremitting psychosis. And they allow us to roam free in the streets.

Then, as I was all tooled up and beside myself with passion over Susan and the Red Sox, the *third* surprise came in a letter from the United States.

"Dear Little Fatty," it began. That was the supposed term of endearment my cousin Lyn coined for me when I was a chubby little boy. That, and other pet epithets like "Neddy Spaghetti with the meatball eyes," or "Neddy, Neddy, two-by-four, couldn't fit through the bathroom door, so he did it on the floor, licked it up and did some more," served to make me self-conscious about my weight my whole life. Nevertheless, that letter was the happiest, best, truest, and most prescient letter I've ever received.

Lyn wrote me about a man she'd met while working in Washington, D.C., whom she'd fallen in love with. She was planning to marry him in December. I wish I still had the letter. I'd never read a letter from Lyn, or from anyone for that matter, that was so unguardedly happy. Lyn could be so caustic and cynical that when she wasn't it brought you to a full stop. This letter didn't have a cynical syllable in it. Just the opposite: The words practically glowed.

Tom Bliss was the man's name. He was a medical student at Georgetown. She'd met him through Kendall Wheeler, a friend of hers at the Washington School for Secretaries. (This was 1967, still the days when a female college graduate like Lyn, Phi Beta Kappa in English from Boston University, would be asked how fast she typed at job interviews.) Kendall's boyfriend, Chandler Van Orman (the names seem straight out of Scott Fitzgerald), played on the same rugby team as Tom Bliss.

Lyn wrote in her letter that she'd never felt like this before. She said it was the way it's supposed to be, the way you dream it might be, the way she'd almost given up hope it ever would be, but now it truly was.

Tom was a really good guy, she was sure I'd like him, but not to worry, we'd still be tight as ever. She couldn't wait for me to get home so she could introduce us. Jamie had already met him and liked him a lot. She loved me and she knew I'd be happy for her.

She was right. I folded up the letter and immediately wrote her a letter telling her, in huge, excessive, overstated caps how much I loved her and how happy I was for her.

While I meant it, I was mighty scared, too. The last person who had told me a marriage wouldn't change anything was my mother before she married Uncle Unger. I felt a bit of panic. I was not at all ready for life without Lyn. But I still had Jamie.

When our group landed in Washington, we said our goodbyes and headed in different directions. I shook Frank's hand and looked for Susan. She had her guitar across her back and was wearing one of the multicolored—red, orange, brown, yellow, green—native garments we'd all purchased.

I was awkward and self-conscious as I tried to hug this woman who was wearing a guitar and an African dress. We did make eye contact long enough for me to see for sure that the soulful glow I had imagined lived within her really did live within her.

And I was able to hug enough of her body to feel really jealous once again of the man she was going home to marry. But Frank had been right. Best to back off.

I wouldn't see Susan again for twenty years. She came into Cambridge and visited me in my condo when I was in my late thirties. The man she'd married—my rival, the mystery man I'd been so jealous of—turned out to be unbalanced. He was mean to her, so she'd finally divorced him, after having four children. It turned out that he had a mental illness, maybe borderline personality disorder, so there was a biological basis to his meanness, Susan told me. He took his own life a year after they divorced. We sat in my living room drinking something with alcohol in it and flirting. I still wanted to jump into bed with her but I still held back, sensing her vulnerability (and also not a hundred percent sure she'd want to), knowing that she'd be leaving town and not wanting to start more than we could finish. We spent the time reminiscing and catching up instead of making love, or maybe that was our version of making love, before we hugged each other and said goodbye once again.

I had no more contact with Susan until we exchanged emails early in 2017. She said how I'd helped her feel like a desirable woman again during that brief Cambridge visit, that later she'd found a really good man whom she married in 1990, instantly making him the father of four. She sent me a photo of the brood of beautiful children she'd begotten with husband number one and the grandchildren they'd brought into the world along the way.

My return from Togo did not doom the Red Sox, thank God. I was relieved that their success had not depended upon my absence. In a final few weeks that were as exciting as any finish in the history of major league baseball, the Red Sox would win the American League pennant by the narrowest of margins, one game, and do it on October 1, the last day of the 1967 season.

The Red Sox luck ran out with that victory. They would lose the World Series in seven games to the St. Louis Cardinals, extending the string to forty-nine years without a world championship. They would come close a couple of times but not actually win the World Series until 2004, eighty-six years after they last won the world championship in 1918.

34.

Senior year at Exeter, 1967–68, was the most transformative year of my life, largely due to the influence of one man: my English teacher, Fred Tremallo (Italian for "three evils," as he, an etymology maven, liked to remind us).

In September, I handed in a short story I wrote which he handed back to me the next day with these words, written in red pen at the end: "Why don't you turn this into a novel?"

As I looked at that comment, that intimidating question, I thought to myself, gee, I always knew Exeter was a tough school, but I never thought I'd have to write a novel.

The more I thought about it, though, the more excited I became. Mr. Tremallo hadn't challenged anyone else to write a novel, as far as I knew.

What made him think I could turn the three-pager I'd handed in into a three-hundred-page novel? Why had he singled me out? I did want to be a writer. That was my number one career goal. Maybe Fred thought I had it in me. I started to mull it over more and more. In fact, I thought of little else.

A couple of days later, I spoke with him after class and said I'd like to give the novel a try. "But how do I do it?" I asked. "How do you write a novel?"

"Scene by scene," Mr. T. replied without even pausing to think. "You know what you want to write about, you laid it out in the story. So just do what we've been discussing in class, focus on scenes, details, characters. Write what you know."

"But how do I organize it?"

"Don't worry about that now," he said. "Just write. Trust your unconscious." It was incredibly freeing advice. Just write. Organize later. And let your unconscious have a hand in doing that as well. "Once your unconscious catches on to the fact that you're doing this," Mr. T. went on, "it will starting working on it round the clock, twenty-four seven. That's the beauty of the unconscious!"

"OK, I'll give it a try."

"One thing, though," Mr. T. added. "You'll have to do this on your own time. This will not be an assignment, so you'll have to do all the assignments the rest of the class is doing. You won't be excused from anything. You're taking on additional work for which you'll get no credit. Do you understand?"

"Yes," I said, wondering what in the world I had gotten myself into.

Each week I added pages, which Mr. T. would read and comment on. At one point he gave me a book by Wayne Booth called *The Rhetoric of Fiction*, which I read as carefully as a rabbi reads the Torah, only, unlike the rabbi, I didn't understand a word of it. But Mr. T. had given it to me, which meant it must contain what he thought I needed to know about writing, so I read it with the utmost dedication, even though I couldn't decipher its messages. Maybe my unconscious did!

Evenings I would often go over to Mr. T.'s apartment—he lived in Wentworth, the dorm next to Bancroft, and we'd talk about life. These sessions meant the world to me. Sometimes other kids were there, sometimes just me. His wife, Ellie, would offer us juice and cookies, and we'd sometimes smoke cigarettes. (This was permitted in a faculty member's apartment or in the designated common rooms.)

We talked about everything under the sun. In French class we were reading Albert Camus's *L'Étranger*, and I was very drawn to its main character, Meursault. The first lines of the novel were *"Aujourd'hui maman est morte. Ou peut-être hier, je ne sais pas"*—"Today my mother died. Or perhaps yesterday, I'm not sure."

In many ways, like Meursault, I attended life, watching, feeling on the outside looking in, not quite sure how to get to the inside, the meat, the main event.

Mr. T. and I talked about existentialism at great length, so much so that I thought philosophy *was* existentialism, that Camus and Sartre defined philosophy. Never mind Plato, Descartes, Nietzsche, or Schopenhauer. I'd barely even heard of them, let alone knew what they said. It was Camus, Sartre, and *"l'absurdité de la vie."*

In that same French class we read Sartre's *Huis Clos* (*No Exit*), which I found incredibly bleak but also compelling. "Hell is other people."

"But the actual line in the play," Mr. T. jumped in when I brought it up to him, "is not 'Hell is other people,' it's *'L'enfer, c'est les autres,'* which is not exactly the same thing. Sartre was getting at something more subtle, the impossibility of knowing yourself except through the distorted mirror of other people. And even if he did mean it the other way," Mr. T. went on, "that's not surprising, because Sartre's childhood was pretty difficult. His father died when he was two, and he was raised by his mother's father and his mother, who was ridiculously intrusive and overly affectionate. Plus, Sartre was short, ugly, and walleyed. Not a happy beginning."

"So you think his bleak philosophy is just because of his unhappy childhood?"

"No," Mr. T. replied. "Sartre was a genius and a great philosopher. I'm just saying his childhood must have colored his outlook on life. Just as your childhood has colored yours, wouldn't you say?"

"I'm sure it has," I said, "I'm just not sure exactly how. My childhood was such a mixed bag of really good and kinda bad."

"You're writing about it well," Mr. T. said.

"Thanks," I replied. He had no idea how much praise from him meant to me. Or maybe he did.

35.

More than any other single event, the war in Vietnam dominated our teens and twenties. We didn't understand why we had to go off to die. Some refused to go; far more just went, obeying the call, doing their duty, serving their country, making the sacrifice, whatever slogan rang true. About fifty-eight thousand of us died there, along with about a million Vietnamese.

It wasn't like World War II, when people faked their age and concealed medical problems so that they could go fight Hitler. Back then, like just about everyone in his generation, my father couldn't wait to go.

For better or worse, we were raised in a country that valued freedom and independence of thought, so when the government told us we should lay down our lives to fight a war that made no sense—no Pearl Harbor, no Hitler, no Nazis—we didn't buy the propaganda. We didn't fear the domino effect: the idea that if the United States didn't prevail in Vietnam, totalitarianism would soon engulf us.

In the 1960s, we began to learn of the whopping lies used to justify sending us off to war, and we were aghast. (I recall the great line from the film *Casablanca*, when the corrupt Captain Louis Renault says, "I am shocked, *shocked* to find there's gambling going on in here!") Waking up to the ways of the world, we protested. We learned why Vietnam was an unjust war, and when we protested, LBJ dug in so stubbornly that he lost the New Hampshire primary to Eugene McCarthy. Promising he would end this war, the idealistic candidate appealed to us kids.

After his embarrassing defeat in New Hampshire, LBJ dropped out of the race. I will never forget the triumph and exhilaration I felt during my senior year at Exeter when LBJ announced on national TV: "I shall not seek, and I will not accept, the nomination of my party for another term as your president."

The Vietnam War ripped through our nation, lit up families like a house fire, and set one generation against the next, in some instances permanently.

My father had rushed to fight in World War II, captained a destroyer escort, fought battles against U-boats, and returned a hero, having given all he had for his country. "All he had," it turns out, was an awful lot. The war cost him his sanity, his marriage, and a daily life with his three boys.

My brother Ben attended the Naval Academy, graduated in 1960, became a pilot and flew jets from, ironically enough, the aircraft carrier *Wasp* before becoming a landing signal officer, the guy who waves in planes as they come in to land if the carrier is listing too much in rough seas. It's a dangerous job but one my brother did well.

Given the family history, then, what my mother said when I told her I didn't want to go to Vietnam should not have surprised or hurt me as much as it did. "So you're a coward," she said. That was unlike her, and it caught me off guard.

I don't remember my reply but I felt completely misunderstood, as so many of my generation did when we were dismissed as cowards, draft dodgers, and un-American. Even the peace sign was ridiculed as resembling the footprint of a chicken.

36.

Like many others, I was caught up in the optimism of the Age of Aquarius, in the notion of universal love, and, naïve as ever, I was starting to envision a Big Change whereby people would stop trying to dominate others but rather come together in peace, harmony, and love.

We were not all high on drugs. I actually hated the way pot made me feel, so I didn't use it, and LSD scared me too much to try. My drugs were alcohol and nicotine. I developed a bad cigarette habit that would take twenty years to break. So it wasn't drugs that made me see the world so rhapsodically.

It was hope, the triumph of hope over thousands of years of human experience. I didn't want to tear down "the system," I was not a political activist. I really just wanted to take advantage of what Harvard had to offer, find a girl who liked me enough to go to bed with me, and find a way to get on with my life.

Still, I did share that vision that we were going to change the basic conditions of life, never mind *how*, we just were going to do it. "You'll see," I'd say to older and wiser people, "a big change is coming."

I smile at myself now, at all of us believers and dreamers back then. But while I smile, it's also true that my work as a doctor and psychiatrist harks back to what most people would call the *craziness* that got going inside of me, inside of millions of us, during those years. It rests on the power of one word: love. It was spoken so much then that it became hackneyed and trivial. It's not spoken so much anymore.

Some took the vision several steps further. Full of righteous rage, acting as if they owned a monopoly on truth, bullying the rest of us with the certitude characteristic of fanatics, they started making demands *or else*. One of my roommates had become active in SDS, while other friends told me I should get off my sorry butt and join the movement.

I had no clue how deep the anger ran with some, on both sides, or to what lengths some would go, until Wednesday, April 9, 1969. Just after spring break, a group of students broke in and physically took over University Hall, the main administration building in Harvard Yard, designed by Charles Bulfinch, class of 1781. My dorm, Thayer, the same dorm Uncle Jimmy had lived in, was catty-corner, about fifty feet away. I had a bird's-eye view.

Not being political by nature, to me it seemed stupid to assault a building and the people in it as a way of protesting the war.

My SDS roommate told me I should join in and take over University Hall, but I didn't see that was worth getting kicked out of college. After I turned him down, he called me a coward. So I had my mother calling me a coward for not going to Vietnam, and my roommate, who'd been my friend all the way back to Exeter, called me a coward for not invading University Hall.

I can still see the students escorting the deans out of that building, actually physically carrying one out, Archie Epps. A female demonstrator got up in his face and said, "You are responsible for killing people in Vietnam." Epps replied, "I am not responsible for killing people in Vietnam. You are using methods here that I thought you objected to—violence and force." The young woman's response: "What the fuck do you know about it?"[1]

---

1. "The Occupation," *Harvard Alumni Bulletin*, April 28, 1969, 22.

Epps was among the first African American deans at Harvard and became my friend decades later. Turns out he'd just returned from his honeymoon a few days before being physically evicted from University Hall.

The protesters' demands, all aimed at getting us out of Vietnam or mitigating Harvard's influence in Cambridge, were nailed on the door of Harvard president Nathan Pusey's house and posted on trees around Harvard Yard: Expel ROTC from Harvard, divest from companies that in any way supported the war, lower rents in buildings owned or controlled by Harvard, and freeze rents in other buildings.

We didn't have to wait long to see what happened, and we should not have been surprised, had we any idea of who Pusey was. A year or so before, he had issued a statement regarding student protesters: "Safe within the sanctuary of an ordered society, dreaming of glory, they play at being revolutionaries and fancy themselves rising to positions of command atop the debris as the structures of society come crashing down." Pusey, apparently feeling that he had to act immediately and decisively because the students in University Hall had unlimited access to all manner of private records concerning faculty and administrators, not to mention a trove of other private material, made what in retrospect was a colossal mistake. He ordered a massive police invasion of Harvard.

With a few hundred other students scattered throughout Harvard Yard, I stood outside my dorm, feet away from the center of the action. At three in the morning, under the cover of darkness, lit only by the nightlights of the Yard, some four hundred troops moved en masse into a place where they did not belong, and yet had been welcomed in by Harvard's president, to perform an action neither they nor we understood. They were paid to do it, we were paying tuition to watch it, and none of it made any sense.

Uniformed state troopers and Cambridge police joined forces as they advanced through the gates across the Yard from University Hall in a terrifying, widening phalanx.

They were dressed in full riot gear, wearing helmets with visors, jodhpurs, boots, armed with pistols, sporting Plexiglas shields and billy clubs, along with mace, tear gas, and God only knows what other armaments. It was overkill in the extreme.

As they got closer, I moved up the steps of Thayer Hall along with my friends. I didn't want to get in the way of a billy club or get maced. We'd been advised to carry wet handkerchiefs in case the police used tear gas, which indeed they did.

The ghoulish image still haunts me of that dimly lit bluish block of troopers and officers, faceless and inhuman behind their helmets, visors, shields, and gloves, marching mechanically, like a vast bubbling blob, pressing down relentlessly, unstoppably upon the gray stone University Hall in the dead of night. Their mission: to confront, overwhelm, and remove the resolute but ragtag, hapless students who were linking arms in a quixotic attempt to prevent this massive police force from squashing them. Foolish, reckless thinking, passionate intensity, and opposite poles of idealism drove both sides, but only one side had physical strength and force.

Soon, police were clubbing and handcuffing students, dragging the intruders out of University Hall, as mayhem escalated while we looked on in horror. The police arrested 122 students.

On the spot, a new cause was born; the students who were clubbed became martyrs. Most of the student body united in opposition to this police action in Harvard Yard.

The psychoanalyst Bruno Bettelheim famously said that we students were working out our Oedipal strivings by attacking the Harvard administration, a defensible interpretation, albeit condescending and incomplete. The injustice of the Vietnam War certainly motivated the students, as well as whatever Oedipal issues were in play.

But what feelings, I wonder, what conflicts was Nathan Pusey working out by sending storm troopers into the heart of the institution he was entrusted to care for?

The following Monday, ten thousand students gathered in Harvard Stadium and voted to close the university down by going on strike. I didn't understand how we could go on strike, since we were not employees but rather paid money to attend the college. Nonetheless, I joined in the demonstration and voted, with almost everyone else, to strike.

During that strike, which would not last long, we met in groups, or we slept late, or both. Not wanting to cease teaching, English professor William Alfred gave a memorable discussion of "The Love Song of J. Alfred Prufrock" under the trees in front of Kirkland House, which I attended.

Alfred was a beloved professor, one whom my brother Johnny had raved about. He was the author of a 1965 Broadway hit, written in blank verse, called *Hogan's Goat*, in which Faye Dunaway got her start. His reading of "Prufrock" was my first exposure to him.

A poet himself, he read the poem almost as if singing. Pausing after bits of reading, he'd offer explication: "Prufrock is a man who is unable to love, that's the heart of this poem." As I would learn, Bill Alfred brought almost everything back to issues of love and loss.

A week after, another large gathering of students convened in Harvard Stadium. We voted to suspend the strike and get on with classes.

In the aftermath, one dean had a stroke and left Harvard, another left for ten years before returning, and twenty-three students were expelled. The terms of ROTC were changed significantly at Harvard, as were several policies regarding governance and community relations.

In a humorous contrast, in reading the history of political activism in colleges, I learned that among the first, if not the first, student protest on a college campus in North America also took place at Harvard. It was in 1766. Despite the fact that the dining hall called the Commons was "the largest and most elaborate culinary establishment in New England," the food served to the students—or the scholars, as they were called—was "as mean and insalubrious as ever, the puddings so hard-boiled they could be kicked."[2] The worst offender, though, was the butter. Imported from Ireland and stored for months, it became rancid. In the so-called Great Butter Rebellion, Asa Dunbar, future grandfather of Henry David Thoreau, caused a disturbance in the Commons by protesting to a tutor the poor quality of the butter. The tutor dismissed him, Dunbar did not take no for an answer, and one incident led to another, including hearings before an admin board, convictions, and suspensions, more hearings, and upper-echelon discussions. The rebellion dragged on for over a week, with some hundred scholars threatening to transfer, perish the thought, to Yale. Dunbar would write an epic poem about the rebellion in which one of the dramatis personae said, "Behold our Butter stinketh and we cannot eat thereof; now give us, we pray thee Butter that stinketh not!"

---

2. Andrew Schlesinger, *Veritas: Harvard College and the American Experience* (New York: Ivan R. Dee, 2007).

37.

The only time I ever saw my father actively psychotic was when I was a sophomore at Harvard. One of my friends, Dave Halvorsen, and I decided to go on a canoeing trip on the Rangeley Lakes in Maine. En route, we stopped to spend a night at my dad's and stepmother's house in Derry, New Hampshire. At the time, Dad was teaching public school, the job he held until he died in 1978.

Dave and I spent a pleasant evening with Dad and Hope, the ghostly, strange woman he'd married after meeting her in a mental hospital. The next morning, when we got up, Dad had left for work, as had Hope. But Dad left me a message written in spit on the mirror in the living room. It read, TAKE THIS MONEY AND HAVE FUN. On the table below the mirror was three hundred dollars in cash.

Two facts tipped me off that something was wrong. First of all, Dad was one of this world's tightest cheapskates. He never would have given me thirty dollars, let alone three hundred, were he in his right mind.

Second was the writing on the mirror in spit. I remembered my mother telling me she could always tell when Dad was about to go crazy because he would start leaving her messages written in spit on the bathroom mirror.

Not knowing what to do, I did nothing. I took the money, and Dave and I went off on our canoeing trip. When I returned to college, Jamie told me that Dad was in the Bedford VA hospital.

The next day I borrowed a friend's car and drove to Bedford, about a half-hour drive from Cambridge. Once I located the hospital and parked, I got out of the car in a daze. I was about to visit my father in a mental hospital. I didn't know what to expect, or how to feel. Part of me felt curious, part of me sad, most of me confused.

I was directed to the locked unit, a term I'd never heard of outside a jail. "Why is it locked?"

"Because the people there can be dangerous and they can try to escape," the woman behind the reception counter told me. "But don't worry, you'll be safe. And your father is getting excellent care."

"Thanks." I appreciated the reassurance that I was not in danger more than that woman knew.

I rang the bell next to the door of the locked unit. It looked like a doorbell, a black button in a brass saucer. After a few minutes, an athletic young man dressed in white opened the door. "I'm here to visit Mr. Hallowell. I am his son."

"Sure. C'mon in. He's doing real good. You can take him out for a walk, if you'd like to."

He was sitting in a chair across the room, smoking a cigarette. Walking toward him, I called, "Dad!"

He stood up and shook my hand.

"How are you doing?" I asked.

"They tell me I'm doing pretty well. I was stupid. I went off my lithium. Stupid."

Soon we were outside, walking around the beautiful grounds of the VA hospital. It was October, the kind of sunny, brisk autumn day New England produces better than anywhere else.

We stood beneath a tree that looked perfect for climbing, with many available branches, when my dad produced a length of rope from one of his pockets. "I'm not supposed to have this, you know. They could take away my privileges. But I want to know, do you think I should hang myself?"

"No!" I said. "Of course not! Why would you ask such a stupid question? Jesus Christ, Dad."

"I'd like to know how it feels to die," he said.

"C'mon, Dad. If you die you won't know what it's like to die because you'll be dead."

"I've always had it in the back of my mind. I once asked Benjie if he'd like me to capsize the boat when we were out sailing in rough seas. Would have drowned us both. He was like you, said hell no, Dad. I asked him why not? He said 'Because I want to live.'"

"Good for him, Dad. I'm glad you didn't capsize the boat. You're not making a lot of sense, you know. Maybe we should go back inside."

Dad kept winding the length of rope around his hand.

"Dad, can you please stop that? This is crazy."

"Well, then, I'm in the right place, aren't I?" Dad said, putting the rope back in his pocket.

We started walking again, and we actually had a normal conversation. I told him about school, how I was majoring in English and taking pre-med courses.

"Don't study too hard like Johnny did. He was so set on getting summa and beating out Uncle Jim that he didn't make time for a social life. Have you joined a final club? I was in the Fly Club, you know. Made lots of great friends there."

"No, I haven't joined a final club. It's different now than when you were there. But maybe I should reconsider. You had fun?"

"Lots of fun. Of course, I had hockey, too. Those four years were some of the best of my life. I hope you make your years there as good as mine. Or better."

"Well, I do have good friends. And I have fallen in love with English. Samuel Johnson, in particular."

"Who's he?"

"Amazing man. He was like a psychologist before there were psychologists. But he was really troubled himself. Depressed, ugly, but a genius. I have this great professor who makes Johnson so real, it's like I know him personally."

"I can see you're excited," Dad said. "That's good."

I was still feeling totally weird, just having talked about Dad's hanging himself and now having this seemingly normal father-son conversation. But I shouldn't have felt weird, as this was pretty much in keeping with my relationship all along with my father. I knew him, yet I didn't know him. He was great with me, taking me to Red Sox games and fishing, but most of the time he was absent and distant. He never talked about Mom, or what had happened. He never talked about my years in Charleston, and he never talked about his years in the war. He did give me advice, though, like not to study too hard as Johnny did, and when I was ten he told me about wet dreams, which felt totally awkward. He was good to me. I just always felt like he held back the important stuff.

We kept chatting, normal father-and-son style, until we got back to the unit. Before I rang the doorbell, I had to ask, "Dad, you're OK? No more thoughts of killing yourself?"

"No, none at all. I'm sorry about that."

"I'm going to have to tell your doctor anyway, OK? Just so I won't worry?" I felt like a tattletale, but I couldn't leave without telling someone.

"Sure, OK. Thanks for being such a good son."

We embraced rather than shaking hands, and I rang the bell.

38.

As I told my dad, I fell in love with the works of Samuel Johnson sophomore year. Not knowing the course would grab and hold on to me as no course ever had before, I innocently signed up sophomore year for The Age of Johnson, taught by Walter Jackson Bate.

Jack Bate was a legend, but at the time I didn't know that. The first day the lecture hall in Emerson 105 was packed. Bate—tall, thin, with tufts of hair sprouting all over his head like white feathers—walked up onto the stage and started to talk, waving his arms a lot. He spoke rapidly and passionately, but also efficiently. "You don't need to waste too much time on James Boswell. If you took this course at Yale, it would be all *about* Boswell, but here, at Harvard, we know better. We will introduce you to the *real* Samuel Johnson. We won't bother with calling him *Doctor* Johnson, as that is pure affectation put on by his admirers. He certainly earned the title, don't misunderstand me, writing the dictionary was a herculean task, but the people who refer to him as Doctor Johnson treat him like a pompous old fossil. He was anything but. He lived a life of constant pain and struggle. Anything he got, he got on his own. He battled two prolonged episodes of deep depression in an era when people didn't understand depression at all. He wrestled with his religious faith, and even though he was a straightforward Anglican, at the end of his life he begged God to forgive and accept his late conversion. He was terrified of death, which probably explains his antipathy toward Milton. In other words, Johnson was a real and complex man, not just the brilliant talker we see in Boswell's biography of him. To be sure, he was the greatest talker in the history of the English language, but he was so much more than that."

And so began my exhilarating ride into the world of Johnson, courtesy of Jack Bate, leading me to write my undergraduate thesis on the religion of Samuel Johnson.

Bate was the lead reader of my thesis and gave me a *summa*, calling it the best undergraduate thesis on Johnson's religion he'd ever read. To this day, that is the accolade of which I am the most proud.

Johnson gripped my imagination as no writer ever had. He had a dark core and yet his life was bursting with love—of people, literature, conversation, underdogs, competition, fellowship, food and drink, walking and exploring, embracing impossibly difficult tasks—in short, of anything and everything his hungry mind and the world around him offered up.

Even though he struggled with depression his entire life and suffered two prolonged bouts of it, and even though he was terrified of death and harbored deep insecurity about his faith in God, and even though he exhibited a host of odd mannerisms and tics to the point that he often ate behind a screen lest others observe his slovenly manners, he triumphed over all of that, becoming one of the great geniuses in the history of English literature and one of the most astute commentators on human nature who ever lived.

Despite all that he suffered, he was, as Bate said, above all else, *sane.*

Having lived with Johnson for a year, at the end of that year, as an English major I had to find someone to be my tutor, just as Johnny had found Bill Alfred and Harry Levin. I took a chance and asked Professor Alfred if he'd be willing to take me on. "Yes," he said, "I'd be glad to."

What luck. For two years I met with Professor Alfred in his study on Athens Street, which was an easy walk from my dorm, Dunster House. Junior year, we focused on plays, and read three hundred of them, about ten a week. Of course we couldn't discuss every one, but I got a large dose of drama, from Euripides to O'Neill.

I never knew who'd be sitting in Alfred's living room when I arrived. One day I couldn't help but take notice of the presence of a beautiful blonde. It was Faye Dunaway, who, having acted in his play on Broadway, had developed a deep personal (though not romantic) relationship with him. Another day, Robert Lowell was there. Alfred called him Cal, as did most of Lowell's friends. Alfred was instrumental in looking out for Lowell's mental health. He said to me, "When the gin hits the lithium, watch out!"

Alfred was friends with so many—with T. S. Eliot, with Gertrude Stein (or "Miss Stein," as he called her), with Lillian Hellman, with Mary McCarthy (whose last name he pronounced "McCarty," I guess in the Irish tradition), with Lowell's second wife, Elizabeth Hardwick, with Elizabeth Bishop, with Robert Frost, with pretty much all the important literary figures of the day. But Alfred could not have been more modest. In our sessions he never boasted or dropped names, but he would tell wonderful stories about all of these people by way of bringing them to life, rather than leaving them as frozen icons in my mind.

A devout Catholic who went to mass every morning, Alfred was the son of an Irish bricklayer from Brooklyn, a man who'd turn up in his Athens Street house every couple of months. Alfred's mother, Mary, had died some years before I met him. He was clearly devoted to her, once telling me the story of how she'd saved up money to buy him a Royal typewriter after, as a ten-year-old, he'd announced he wanted to become a writer. "She lugged it on her hip almost a mile, and it weighed a ton." He took such good care of his father, I believe, in part out of loyalty to her.

Now and then he'd invite me to go "slam a gate," which meant go to the Parthenon, a Greek restaurant a few blocks down Mass

Ave toward Central Square, or, if other people were joining us, to the Athens Olympia in downtown Boston. Happily, I loved Greek food. Alfred, true to his Irish roots, would always order meat and spuds, the Greek version being lamb and roast potatoes. Both restaurants had white linen tablecloths on top of pads, so there was a cushioned, luxurious feel as you rested your arms on the table.

The dinners always began with martinis and appetizers. I loved taramasalata, and both restaurants made it especially well. It's hard to find good tara, as I called it. But the martinis stole the show. Mr. Alfred and I usually had at least three apiece. By the time the main course came, we were very well lubricated, talking up a storm, laughing about faculty foibles, politics, the Irish ("You need to know two things about the Irish," he'd say; "first, they are basically insane, and second, they harbor a fundamental hatred of life"), campus protests ("Pusey was a fool," he said, "but the protesters acted like moral jocks, as if they and they alone knew the truth"), education ("The worst thing Harvard or any college can do is fill students full of stupid wonder"), T. S. Eliot ("He could be quite catty—one day we were being driven in a car and I told him we had our mutual friend Jonathan Griffin's new play in the trunk, to which Eliot replied, 'Well, then, we should have a very smooth ride'"), and love ("You can't explain falling in love, it's just if you like the cut of her jib"). By the time we waltzed our way home, we'd slammed a gate for sure.

Alfred dressed nattily, but in order to have more money to give away, he bought all his clothes used, at Keezer's, a legendary Cambridge consignment store near his house where JFK and many others used to sell their old clothes.

My senior year, he took in a homeless teen who'd been in trouble with the law. Sometimes he'd call me in the middle of the night to drive him to a jail "to bail out the boy." Alfred did not have a car himself, nor could he drive.

His faith in God was unshakable. One day during our session he pointed to the Franklin stove in his study and said, "I am as sure there is a God as I'm sure that stove is sitting there." But he was anything but pious. He hated the religious hypocrites who sullied religion. At the same time, he would never write anything in his plays that might offend a priest.

When B. F. Skinner's book *Beyond Freedom and Dignity* came out in 1971, Alfred said to me, "Beyond freedom and dignity? That's what B. F. Skinner is." Alfred took concepts such as freedom and dignity seriously, and arguments against them personally, having seen people die to protect them when he served in the Army tank corps during World War II.

At one point during a tutorial, we got to talking about my plans after graduation. I was thinking of medical school, but also wondering about graduate school. "Maybe I should go on and get a Ph.D. in English," I said.

Without pausing even a second, Alfred instantly responded, "Oh, no, don't do that, you'll end up hating books!" So rare, a teacher counseling a student *not* to do what he had done himself, not to follow in the steps of the mentor. "Become a doctor and a writer. Medical training is great training if you want to write about life. Lots of great writers were also doctors. Keats, Chekhov, Maugham, Williams, to name a few. You're much more a doctor than an academic," he said.

I was so grateful to him for giving me that advice. If he had urged me to go on and get a Ph.D., I might well have done it, if only because that would have been by far the easier route for me.

Toward the end of junior year, I had to decide on a topic for my undergraduate honors thesis. Alfred and I batted around some ideas: maybe write about *A Long Day's Journey into Night*, because we both loved O'Neill and the play reminded me so much of my own family; maybe write about T. S. Eliot's literary criticism and tap into Alfred's personal knowledge of Eliot; maybe write something about the theory of comedy, as Harry Levin's course on that had really turned me on; or write something about my favorite novel, *The Great Gatsby*. But again and again, I kept coming back to Johnson, whom Alfred also loved. "The only problem with writing about Johnson," he said, "is that it's hard to write about someone who's said it all already himself."

"I guess so," I replied, "but I'd like to try."

"Well, then you might look at his religion," Alfred said. "Get into his psychology. Start with that great line in *The Vanity of Human Wishes*, 'the secret ambush of a specious prayer.'"

So I was off to the races.

I would delay a year, because, after waiting tables at Pate's Restaurant in Chatham at night during the summer and tutoring high school kids in math during the day, I'd saved enough money by mid-October to take a year off and go to London.

Waiting tables itself had been an adventure, learning how to carry trays laden with lobsters on the palm of one hand held high above my shoulder, making friends with Joe Kublicki, a big, tough guy who showed me the ropes, arguing with the racist bartender who insisted blacks were such good athletes because they had "an extra tendon, like monkeys," waiting on customers who got so drunk they'd either leave without paying the check, and I'd have to chase them out into the parking lot, or they'd leave a tip so huge I'd have to ask them if they really meant to do that (otherwise they'd only cancel it the next day), and then, after work on Mondays, going down the road to another bar and watching the new show *Monday Night Football* on TV, being entertained by Howard Cosell, Dandy

Don Meredith (who'd spar with Cosell), and the classy Frank Gifford, who'd mediate between them.

Once in London, I rented a flat just over the Hammersmith Bridge in Barnes from a flamboyant playwright named Paddy and her not flamboyant barrister husband. Paddy would talk to me about writing and tell me to be sure to have a job that would generate income so as not to have to rely on writing to pay the rent. Or marry someone who had such a job. For such a flamboyant playwright, Paddy was also practical.

I fell in with a circle of writers, including my friend from Exeter and college Jon Galassi, who was at Cambridge on a fellowship (and went on to run Farrar, Straus & Giroux and become one of the country's premier editors as well as a poet and novelist in his own right), the poet Judith Thurman, whom I'd met through Bill Alfred (Judith went on to write a biography of Isak Dinesen, which was turned into the movie *Out of Africa*, and one of Colette, and to write regularly for the *New Yorker*), Jonathan Griffin, who wrote that weighty play and was an excellent poet, and another playwright, David Pinner, a rotund Falstaffian figure whom I came to adore. We would meet almost weekly for parties, readings, dinners, and general good times. Judith and David I'd see almost every day.

My brother had given me an introduction to the chemist and novelist C. P. Snow. I will never forget meeting in his extremely spare sitting room in London. (Was the spareness an expression of left-wing politics? I didn't know.) We talked. Lady Pamela and Lord Snow were congenial, but I was never invited back—maybe because I wasn't my brother Johnny, or I had not yet read any of Lady Pamela's novels, or both.

While in London I fell in love with a wonderful woman. She was the epitome of style: elegant, witty, and beautiful. She wasn't trying to cast a spell but I fell under it nonetheless.

It was romantic love at its most blind. Not that she wasn't worthy of being adored, she absolutely was. But even though she gave me no encouragement in the physical realm of our relationship—we never even made out, never mind going to bed together—I summoned up my nerve, or lost my mind, and asked her to marry me. She gracefully, tactfully, and ever so gently declined. That she did so without saying "Are you completely crazy?" testifies to her immense kindness.

After I returned to the United States we kept up a correspondence that gradually petered out. Around that time, I learned from Jamie, who knew her as well, that she was sexually more interested in women than in men.

I suppose she might have told me before I more or less made a fool of myself, but on the other hand, I understand why she didn't. She wasn't leading me on sexually in the least—why should she tell me she preferred women to men as romantic partners? Whatever, I treasure the time we spent together and feel nothing but warmth toward her now. I also, once again, shudder at my naïveté.

With the money I had left, before returning to the States I took a trip by myself to Greece. The nights I spent on Mykonos would have been perfect had I been with a woman. Being only with myself, they were depressing. A healthy young single male depressed in Mykonos? Go figure.

I cut my trip short and took the Orient Express from Athens through Belgrade, Vienna, Innsbruck, Zurich, Paris, and Calais and thence back to London. I almost didn't make it, though, because I got off the train in Belgrade to get a snack and a cocktail. When I reboarded, like an idiot I got on the wrong train.

My passport, my luggage, my traveler's checks, *everything*, was on the other train. Here I was in a Communist country in 1972 with no ID, no money, and no language other than English and a little

French. Mercifully, God smiled on me and directed me to a conductor who spoke English. He told me to get off at the next stop and wait. "Be sure to get on the very next train that comes into the station, heading in the same direction as this one. That's your train." I wanted to hug him.

As I stood on the barren concrete platform in a small station in Yugoslavia, I wondered, what if he was playing a trick on the stupid American, or if he'd misunderstood my question? What if I'd misunderstood his instructions? I watched stray papers blow across the track and felt a chill as I prayed that my train would soon come.

Those fifteen minutes were the longest of my life. I paced. I kicked an empty Coke can that rolled my way out of nowhere. I chain-smoked six or seven cigarettes. Finally, the next train pulled into the station. I jumped onto it like a life raft and frantically started looking for my compartment.

Within a minute I found the group I'd been sitting with, including the European mother figure who immediately started scolding me as if I were her son. "Where have you been? We were worried sick! You left everything here on the train! Do you know how much danger you were in? You could have been arrested and left to rot in jail! Do you even have a brain? You have no idea what you put all of us through."

I could have kissed her. She was so concerned, so maternal. I promised her I'd learned my lesson and would never, ever be so stupid again.

"Good," she said. "You could have given me a heart attack." The other members of the compartment were smiling. It was wonderful to be so looked out for by strangers.

39.

Why Cousin Lyn and Tom Bliss picked the day before New Year's Eve to get married is beyond me, but it seemed fitting that Lyn would not choose just any day for her wedding. In a more symbolic sense, their wedding marked the start not only of a new year, but new lives. Lyn would say to me years later, "My life is divided into two parts: before Tom and after Tom."

I had a ball at their wedding, drinking a lot of beer. When it came time for toasts, I stood up on the banquette where I was sitting and raised my glass.

I was expansively emotional, the way I was (and am) when I drink too much, but it was all heartfelt. I was (and am) nothing if not sincere, transparently so, often too much for some people. What I didn't know at the time was that Tom really does not like big displays of emotion; they make him feel awkward, especially when directed his way. Tom must have wanted to crawl under the table as I went on and on about how great he was and how glad I was to have this new brother. He was a good sport, though, as always. It was in his DNA.

He also was under a permanent spell cast by Lyn. He believed he'd just won the Irish Sweepstakes. Best of all, so did Lyn. They'd each found the person of their dreams. I know—it never happens in real life; but it had happened to them.

As Lyn had written me earlier when I was in Africa, her friend Kendall had suggested they go watch a rugby match Kendall's boyfriend was playing in. That's when Tom Bliss caught Lyn's eye. She liked athletic guys, and he was that, for sure: fast, quick, and a cagey runner. (I know because I went to see him play rugby a couple of times, and years later, when we'd play games in the yard or on the beach, he would run circles around me.) He was a star on the field, although he would deny that. A Jimmy Stewart kind of guy, he'd deny being a star at anything, even though he was good at almost everything he did. And, in part because of a prominent nose, he was drop-dead handsome.

They went out for drinks after the rugby match, and, to hear Tom tell it, that was all it took. He knew Lyn was the one. "The minute I saw her, I couldn't believe my luck," he'd tell me years later. "I thought she was way out of my league, but for some reason she seemed interested, so what the hell, I kept coming around."

Dr. Zhivago had just hit the screens, so Duckie dubbed "Lara's Theme" Lyn and Tom's song. Lyn and Tom thought having a song of their own was way too hokey, but I bet underneath all their sophistication (and after a few drinks), they'd feel it when "Lara's Theme" came on, wherever they were. For my mother and father, their song was "Smoke Gets in Your Eyes." Long after the divorce, whenever that song came on, I saw it affecting Mom more than a little.

Following Lyn's adolescence in which small-town teenage boys laughed at her after planting a crazed, shit-shooting turkey in her car and an English teacher took advantage of her, and an early adulthood in which she fell for an older man and tried to make believe that that relationship was ideal, now, with Tom, the dream actually did come true.

Lyn opened up this Catholic boy from Westfield, New Jersey, who'd grown up in a family that by anybody's standards (except maybe their own) was normal. Tom was the oldest of five children. The closest thing to a hang-up that Lyn, ever the psychological detective, could decipher from Tom's childhood was this: His dad went off to war when Tom was an infant, giving young Tom his mom's total attention. Then, when his dad came back from the service, little Tom got displaced by his father, leading the boy Tom, and then the man Tom, to be inordinately possessive and jealous of whatever woman he was with. Other than that quality, which isn't that unusual, Tom was about as well adjusted as a person could be. Still, never daunted, but unable to find any imperfections Lyn could use against this man in the heat of battle, she pulled out the Hallowells' worst insult: *boring.* While Tom wasn't boring in the least, that was the best Lyn could come up with when she needed *some* put-down to throw at him.

But boring was *so* what Lyn, and all the rest of my family, needed. Stable. Reliable. Consistent. We'd never had that. Hello, boring! We thought we'd never findja!

Lyn did show Tom a different way than what he'd been used to. One night, for example, she dared him to take a letter she needed to have mailed and go to the mailbox around the corner from their apartment. Sort of like when Lyn asked me to go to HoJo's and get her a cheeseburger, Tom asked why she couldn't go mail her letter herself.

With Tom, she had inducements she couldn't use with me, and he was quite susceptible to them, so he agreed, OK, he'd take her damn letter and mail it for her.

Oh, good, she said, except there's one additional requirement if you want to get your reward.

What might that be? inquired Tom.

You have to perform this mission without clothes. You have to go out stark naked. As a jaybird. Without a stitch.

No way. There was no way this nice Catholic boy from New Jersey was going to risk getting arrested for indecent exposure, get kicked out of medical school, and forfeit his dream of becoming a doctor on some kind of a crazy dare.

I have no idea how Lyn talked him into doing it. It was her genius. Once she had you, she really had you. She could talk a person into doing almost *anything*.

Tom took off his clothes and used all of his athletic skills to run and dodge and hide behind bushes around the block to the mailbox, drop the freakin' letter in the slot, then, always on the alert for flashing lights or any sign of police, hustle and dodge his way back to the safety of the apartment. When he got there, he faced one last barrier. Not having a key—he was naked, after all—he had to knock on the door. Lyn, of course, couldn't resist keeping him waiting just long enough to scare the daylights out of him. When he barked in the loudest whisper he could muster, *"Someone's getting off the elevator! Open the door now!"* she relented and let him in.

Their wedding was picture perfect. Tom had a bunch of handsome ushers and Lyn a bunch of beautiful bridesmaids. The hidden agenda was that Johnny, who was an usher, would ask out Avery, who was a bridesmaid. That this would not happen because Johnny was gay—a secret known only to a few people, not including me— did not prevent a romance from blossoming at the wedding, though. My brother Ben fell for Avery, and she for him. They drove off to Boston after the wedding and have been together ever since.

40.

About nine months after they got married, Lyn and Tom had their first child, a girl they named Mary Josselyn, Molly for short.

Tom had graduated from Georgetown Medical School the previous summer and was doing his internship at Rhode Island Hospital in Providence, where he would also go on to do his residency in Orthopedic Surgery.

At the time, I was a freshman at Harvard, Jamie was a senior. We could easily drive down to Providence for dinner or to keep Lyn company when Tom was on call.

Back then, Lyn had been urging me to go to medical school because Tom was a doctor, she thought I'd make a good doctor, she wanted me to be practical and find a way to make a living (which others in my family had not done terribly well), and she thought being a doctor would make me happy. She wanted everyone to be happy. This was the major ambition she took out of childhood: a burning passion that the people around her, especially her family, would be happy. This was her ruling purpose, her daily preoccupation. To that end, she could be what's often called controlling and manipulative, but it was in the service of trying to help us lead happy lives. I, for one, really needed the guidance she provided. But sometimes she took it a tad too far.

For example, one day when I came down from Cambridge to visit and spend the night, along with Jamie, she decided it would be a good idea to straighten my hair. I had naturally curly hair, which

I had inherited from my mother. For some reason, Lyn seemed to think I'd look better with straight hair.

She'd always been able to talk me into anything and, sure enough, she talked me into this ridiculous undertaking. The process of straightening took hours and was a real pain in the ass. Once it was done, I looked OK with straight hair, but certainly not better enough to make it worth getting regular treatments to keep my hair straight. That experiment ended.

But this was who she was. She loved keeping all of us off balance. She never allowed herself to become predictable. Like being boring, being predictable was a sin. One evening she contacted a girl I was going out with and persuaded her to come to Providence and jump out of a big cardboard box as a surprise present to me. The girl was a good sport, but our relationship ended soon thereafter.

But the bigger picture she was creating inspired me. Seeing the happy family she was growing child after child with Tom proved to me it could be done. After Molly, they had another baby, named Thomas, after Tom, but called Tim. Not too long after came James, named after Lyn's dad as well as Jamie, but called Jake. Lyn had wanted a girl this time, so when she got Jake she persuaded Tom to adopt a girl from Korea, whom they named Anna. Not quite finished, Lyn got pregnant one last time and had Ned, named after me. Between Molly and Ned there are twelve years; five children born between 1968 and 1980.

Before all the kids arrived, we used to play various games to stay entertained. Not just Lyn, none of us did boredom well. Lyn invented one game that was especially diabolical. Tom, Lyn, Jamie, and I would sit around a table with a glass in the middle of it. Someone would take a single layer of a napkin, wet the rim of the glass, and lay the napkin atop the glass so that it stuck, and then place a dime on top of the napkin.

This was back when both Jamie and I smoked cigarettes. One of us would light a cigarette. Each person then had to burn a bit of the napkin with the cigarette, enough so that you could see a portion of the paper burn. Whoever made the dime fall into the glass lost the game.

Here's where it got interesting. The loser had to make a phone call to whomever the others designated. Since we knew each other well, we could designate the person the loser would be most embarrassed to call. For example, Lyn would have to call one of the boys who put the turkey into her car.

Predictably enough, we were usually drinking when we played this game, which made steadying the cigarette more difficult but making the phone call less of a trial. One time I had to call Uncle Unger and I actually got him on the phone. I don't remember what we talked about, but I know the conversation was brief.

The funniest time, though, was when Tom lost. Lyn made him call her old boyfriend, Dick. For all that Lyn had loosened up Tom enough to marry her and go along with her antics, his DNA was still that of a reserved, somewhat uptight, respectable man. He was not the type to call old boyfriends, even on a dare.

Watching him call Dick was hysterically funny. I can still see him, receiver to his ear, closing his eyes praying that Dick would not answer the phone.

But Dick did answer. There was a slight pause as it hit Tom that he had to speak. He then cleared his throat. I'll always admire him for not just hanging up and saying "Fuck this." Instead, likely praying it was a wrong number, he said, "Hi, is this Dick?"

When the reply came in the affirmative, Tom cleared his throat again. "Hi, Dick, this is Tom Bliss." Awkward doesn't begin to describe what I am sure Tom was feeling. He looked sick.

Tom listened to Dick's reply.

"Well, the reason I am calling you is that we are playing a game, and I lost."

We all heard the click as Dick abruptly hung up.

I suppose it was a sadistic game but it made us all laugh. People's feelings could have gotten hurt, but they never were, at least as far as I knew.

When she wasn't designing diabolical games or straightening hair, Lyn was helping to manage the lives of Tom and her children and the people she loved, her mother and father, Jamie and me, and the odd cast of characters she'd made friends with over the years.

Gammy Hallowell and Duckie, extremely extroverted and comfortable with the Social Register set, intimidated Lyn. Born with her father's anxiety gene, she never developed their social confidence. She got a master's in Arts and Teaching but was never able to hold a teaching job, largely because anxiety got in the way. Instead she worked behind the scenes, where she felt confident, coaching her kids, Jamie, and various others whom she took on informally. I was just one of the many to benefit from her ongoing counsel.

As much as Tom stabilized her, which he did beautifully, Lyn also helped him with his various social anxieties. On their way to cocktail or dinner parties, Lyn would coach Tom on topics of conversation and how to stand with a drink without seeming to be too tongue-tied or anxious. As much as she credited Tom for helping her, he also credited her for helping him.

We developed a rhythm that lasted years. Jamie and I would drive down and help them move, first from one apartment in

Providence to another, and then to the house in Rehoboth, Massachusetts, some ten miles east of Providence, where they lived once Tom finished his training and they could afford a house. With a swimming pool, no less. And a huge yard. It would be their only house, where all the kids grew up.

Lyn loved to garden, as did Jamie, and soon so did Tom. I never became the gardener they did, but the grounds around the house in Rehoboth could have been shown in a magazine or on a garden club tour.

Lyn didn't stop there. Perhaps influenced by her early years on Cloverluck Farm, she populated the Rehoboth property with a wide variety of animals. First, there was a horse that lived in the barn next to the main house. Molly developed a passion for riding, so there was a good reason for the horse.

But Lyn added an array of other critters: dogs and cats, of course, but also guinea hens, ducks (they had a small pond), geese, peacocks, and turkeys. When you drove up to the house you were greeted by a god-awful honking cacophony dominated by the guinea hens and peacocks, but with all the other animals, except the cats, joining in.

Through the years, Jamie and I went to Rehoboth for Thanksgiving and Christmas. Traditions developed. Lyn always made the Brandy Alexander pie for Christmas that Tom's mother used to make and Tom loved. And she'd make a clam dip with sour cream and cream cheese that Tom also loved. She'd produce a Christmas card every year with all the kids on it, designed to be more funny than show-offy, and she'd somehow manage to mail them all out before Christmas while riding herd on each child's education.

As a surgeon, Tom earned enough to send the kids to private schools, all eventually ending up at Moses Brown in Providence, where they not only got a top-notch education but came as close to spiritual or religious exposure as they ever would, as Moses Brown adhered to Quaker traditions.

Given our family's Quaker roots, Lyn could embrace this. Tom, who, with Lyn's encouragement, had decided Catholicism, the tradition into which he was born, made no sense, and was happy with the Quaker influence. Family members sometimes even called each other thee, and they opened each Thanksgiving and Christmas dinner with a moment of silence while holding hands around the table.

As the family grew, Christmases grew into increasingly lavish productions. Mostly, they were fun. But Lyn orchestrated them with such care and a critical eye that I sometimes felt as if each gift I bought had to pass her inspection. Indeed, she would ridicule gifts she thought did not demonstrate enough thought on the part of the giver. (I was often the recipient of such ridicule.) Lyn would also insist that each gift be opened by the recipient with enough time for everyone to see the gift and comment.

When there was just Molly, and the adults, this was fine. But when there were five children, not to mention the additional adults Lyn brought in along the way, the opening of presents became all but interminable. Bloody Marys helped.

The years run together, but I can see us all opening presents, nibbling clam dip, and drinking beer, wine, and cocktails, ending up seated around the long antique dining room table eating turkey, stuffing, and Jamie's homemade cranberry sauce followed by Brandy Alexander pie.

I remember, one Christmas in my thirties, wondering how long this could last. I imagined each Christmas as a red light on a horizontal white bar, and mulled over how many red lights I'd manage to see before the red lights went out. At sixty-seven, I'm still counting.

41.

My senior year at college, I had to apply to medical schools and write my honors thesis. It was a highly competitive time to try to get into medical schools, and while my GPA was high, my score on the MCAT, the standardized test you take to get into medical school, was not high enough to make my acceptance the slam dunk it otherwise might have been.

I regretted not having taken chemistry and physics at Exeter. I'd taken plenty of math and done well, but in science I lacked a strong foundation. I'm pretty sure that if I had taken science courses at Exeter, I would have done much better on the MCAT, but at Exeter I didn't know I would end up applying to medical school. At the time, I wanted to be a writer or an English teacher.

Even though I ended up graduating magna cum laude with a summa thesis and outstanding recommendations, I did not get into a single medical school. This was a huge disappointment and a profound embarrassment. I had taken all the pre-med courses and done well, I had excelled in all the other courses I took, I had put up an excellent record, but because my MCAT was not high enough, I didn't get in anywhere.

Part of me wanted to say *Screw it, if med school doesn't want me, I don't want med school.* But then I talked to Lyn and Tom and the med school adviser. I took a deep breath and applied for a job as a research assistant with Dr. George Blackburn.

I spent a year working for George. The idea was that if I did research and demonstrated a strong desire to go to medical school, I might get in the second time around.

George was an unforgettable character, demonstrating more energy, drive, and dedication to his work, and to helping others, than just about anyone else I've ever met. Both a surgeon and a Ph.D. in biochemistry, he did research into nutrition, particularly the nutritional care of the surgical patient, following in the footsteps of the great Franny Moore.

His labs were at the old Boston City Hospital, near the Fifth Harvard Surgical Service, so I got to attend surgical grand rounds with some of the greats—John Mannick, George Clowes, Bill McDermott. George had a team of about a dozen: nurses, would-be doctors like me, surgical fellows from the UK or Europe, grad students, other docs doing research.

George was a brilliant dynamo, nonstop action, nonstop ideas, nonstop demands. We were always preparing one grant proposal or another, tracking down one reference or another, running some errand or another, and trying to keep up with everything else he wanted us to do. We were always on red-alert.

Looking back, I can see that George had both dyslexia and ADHD. Left-handed, like many of us with those conditions, when he couldn't think of a word he simply substituted "thing." So a sentence could go, "Ned, I want you to go up to the ward and take the, uh, thing . . ." Somehow I learned how to fill in the blanks.

One of my jobs was to take the calipers Bruce Bistrian would give me, go to all the patients on the surgical ward, and measure their triceps skin-fold thickness as well as arm circumference. This scared the patients because many thought I was measuring them for their coffin! I had to explain to them in considerable detail that what I was doing had nothing to do with their demise, only their nutritional status. Once they understood, they were quite happy, even proud, to be measured.

Then I had to record their weight and serum albumin level from the chart, gather up their twenty-four-hour urine and fecal collection, put those goodies into plastic bottles with preservatives, transport them across town, gather data from the previous day, enter it into the computer, run the program, and graph the results. I did this day after day.

We had a hardworking, fun-loving team. We often went out drinking after work (not with George) and we came to really like each other. Then and there I got hooked on teamwork.

I also became smitten with a nurse on the team named Julie.

Before I knew it, we were together all the time. Julie was one of those rare people who are good to the core. But she was also playful, full of laughter, cute as could be, and very smart. With me she was a really good sport, putting up with my sometimes ridiculous, impulsive behavior. For example, late one wintry night we were driving down Mount Auburn Street in Cambridge when we passed an old man standing in a snowbank. I stopped the car.

"Why are you stopping?" Julie asked.

"That man back there, he looks stuck." I backed up the car, put on my flashers, and got out. Sure enough, the man was stuck. Julie and I helped fish him out of the snowbank, only to find his right leg was amputated below the knee. His prosthesis was pretty grungy. When we asked him where he lived, he told us nowhere. He was homeless.

I said we can't just leave him here, and Julie said OK, let's take him to a shelter, and I said can't we take him home and clean him up, and she amazingly said OK, we could take him home. I think I've developed this habit of taking people in because Duckie and Uncle Jimmy took me in. In any case, it was not me but Julie who took on the lion's share of the work when we got this fellow back to the apartment.

He stank—no other way to put it—so badly that it was nauseating to be in the same room with him. Julie took him into the bathroom, drew a tub of hot water, took off his clothes, and helped him into the tub. I helped ease him into the bath as well, but Julie ran the show, knowing what she was doing far better than I did.

The tub quickly filled with such slime that we had to drain it and fill it up again. We must have used a full cake of soap on the poor man, who was happy as a pig in poop, moaning with pleasure and smiling from ear to ear as he was bathed by this lovely nurse.

By the time Julie finished washing him stem to stern, he actually pinked up, smelled like soap, and was clean. He spent the night on our couch.

"Please do me a favor," Julie said to me as we were falling asleep. "Don't fish any more men out of snowbanks, OK?"

We didn't have a lot of money, but between her job as a nurse working for George and my stipend as a research assistant, we had enough money to have fun, and I could still send some to my mother. We loved going out to eat; in Cambridge and nearby, there were lots of cheap places to choose from. We also loved ball games, cheering on the Celtics and Bruins, and we loved movies.

The research panned out well. Our data showed that the surgical patients at Boston City Hospital came into the hospital in better nutritional shape than when they were discharged. That was a dramatic finding, because declining nutritional status impairs recovery big time, especially wound healing.

The paper was published in the *Journal of the American Medical Association* and became one of the most cited papers of the decade.

George gave me various other unsavory jobs, like decapitating rats in a guillotine and "harvesting" their blood in a paper cup while they exsanguinated, taking the blood over to MIT, and entering the data into Prophet, the software system I was trained to use; weighing people at the obesity clinic he ran; editing and rewriting his incredibly bad prose; and writing editorials that were above my head but nonetheless gave me good experience. All these jobs contributed to the cause, George's mission: improving the nutritional care of the surgical patient and raising awareness of nutrition in general.

When I got into Tulane Medical School, Julie and I had to make up our minds what to do. We decided I would go off to New Orleans and she would stay behind.

We went to see *The Way We Were* with Robert Redford and Barbra Streisand and cried. The theme song from that movie became our touchstone.

We could have moved together to New Orleans. Julie could have worked as a nurse while I was a med student. But we were young, and moving to New Orleans together would have been tantamount to marriage. As much as I did love Julie, I wasn't ready.

She is now happily married, as am I, so I think we made the right decision. George Blackburn died in 2017 after an illustrious career at Harvard Medical School. In addition to mentoring people who wanted to go to medical school, like me, he made a huge contribution to the field of nutrition.

42.

In August 1974 I boarded a train at the Back Bay Station in Boston, heading for New Orleans. Jamie drove me and my two big suitcases, plus a smaller bag, to the station. He waited with me on the platform for the train to what seemed like Neverland.

Standing with Jamie, I felt mostly afraid. Excited, yes, I'd worked really hard to get into medical school, having had to apply twice, but now that it was upon me, I wondered if I'd done the right thing. The last time I'd boarded a train to go south, it hadn't turned out so well.

Jamie made small talk, but I could tell he was sad as well. He would miss me, I would miss him. We'd gone out for "a few beers" as we called them so many times, and now that would end, along with so much else.

I felt like saying *Why don't I just throw out this med school idea? It was a long shot anyway. I should be an English teacher or go to law school, hit something in my strike zone. Med school is such a reach. I oughta stay here with you and Julie and Tom and Lyn . . .* The thought of Lyn brought me up short. She'd kill me if I backed out now. My lot was cast.

The train pulled in. Jamie helped me lug my bags onboard and find a seat. I couldn't afford a sleeping compartment, so I was going to sit up all the way.

We were not big huggers, but we were at that moment. I was going to miss him so much.

I waved to him through the window as the train slowly left Back Bay Station. I had no idea what the next four years would bring. I sat back in my seat and said a prayer.

Some thirty-six hours later, the train pulled into New Orleans. I had found no companionship on this trip, and when I arrived I was feeling pretty lonely.

I caught a cab and piled my bags into it. The dorm where I'd live over the next few years was a made-over high-rise hotel. My "apartment" consisted of a tiny kitchen, a sitting area with a desk, and a bedroom. It was about as sterile as it could be, and the minute I walked in I wanted to turn around and leave. Existential crisis, hello!

Looking out one of the windows, however, I saw a vertical sign that read, from top to bottom, J - O - Y. Talk about a message from beyond. Not only that, but the sign was the marquee of a movie theater. Heaven!

I quickly unpacked all that I needed to unpack and made a bee line for JOY. When I got there, I saw right away that this theater had seen better days. No problem! Fancy I did not need. A movie I very much did.

Over the next three days I saw *The Sting* eight times. Movies have always been my go-to escape hatch.

Then the day arrived for med school to start. Dr. Walsh, the Dean, welcomed us first-year students with words that warmed my heart. "Now that you've been admitted, it is almost impossible to flunk out." It wasn't true, some did flunk out, but it was enormously heartening to know the Dean was rooting for our success.

We were each given a long aluminum box of bones, all the bones in the human body to a box, to take home with us so we could learn about each bone, hold it, and get to know it. I would sit, tapping a humerus on my knee, as I read the anatomy text.

We had to learn the origin, insertion, the nerve that connected to it, and the action of every muscle in the body. There are plenty of them. I made a flashcard for every muscle, the name of the muscle on one side, the origin, insertion, nerve attached, and action on the other side. It was tedious, but for me there was no other effective way.

Memorization ruled. There were classic mnemonics for certain topics we had to memorize, like the twelve cranial nerves. Rather than use the mnemonic med students had been using for scores of years—"On Old Olympus' Tipmost Top, A Finn And German Viewed Some Hops"—I made up my own: "Old Options Occur To Those Of Maximum Means About Finding Very Glamorous Valets And Whores." It worked, except I had to tell myself whore began with an *h* for hypoglossal, not a *w*. Mine was actually better than the traditional one because with "Of Maximum Means" I was able to include cues for the three divisions of the trigeminal nerve, the ophthalmic, mandibular, and maxillary.

This was the first two years of med school: Dissect and memorize. Be it gross anatomy, histology, neuroanatomy, embryology, biochemistry, it was all about memorization. We tried to take inspiration from the knowledge that we were learning fascinating facts about the human body, facts we would need to have at our fingertips to be good doctors, but it was also hard because we knew we would forget most of it, that doctors retained only what they used. Not many doctors still know the Krebs cycle they once memorized. It's more a rite of passage than anything else.

But pass I did, thanks in large part to dinner most nights with my classmate, Steve Bishop. He had been raised Catholic and had thought long and hard about God. We had long talks about religion, as well as literature, philosophy, ethics, psychology—all the stuff we missed from our college days. We'd go to a cheap cafeteria, eat really good food (it's hard to find bad food in New Orleans), and talk and talk and talk. Steve was a brilliant guy. I never would have been able to tolerate those first two years of med school had it not been for him.

I found ways to stay interested, and one time almost got into big trouble for doing so. We started the year by taking a class in gross anatomy. We worked on our cadavers in groups of six. Each cadaver was stored in a large aluminum tank the size of an extra-large coffin, but on stilts. We'd open the tank using levers, raising the cadaver so we could dissect it. At the end of each session, we'd lower the body back into the tank and close it up for the night.

The cadavers reeked of formaldehyde and so, naturally, did we. Once we all got over our initial shock at seeing a stiff, dead body, and set ourselves to the task of exposing and identifying every muscle, nerve, vein, artery, lymph vessel, organ, node, lobe, and bone, we quickly forgot that the entity we were so carefully slicing into was once a living person, as alive as any of us.

It's just as well that we didn't keep that fact in mind because you wouldn't want to do to a human what we did to those cadavers. We didn't disrespect them, in the ways you see in B movies about medical students, but the process of dissection is a euphemism for carefully cutting a body to shreds. By the end of the term, the cadavers were barely recognizable as human, having had the skin peeled back, every organ sliced into, the skull sawed open, and every smallest vessel or fiber laid out for identification. The generous people who donated their bodies to the medical school did us a huge favor, but I am quite sure each and every one of them would be grateful that they did not have to watch themselves being dissected.

I really liked the people on my dissection team. Our leader, by acclaim, was a local boy, Clayton Griffin, about six foot five, and I'd guess around 350 pounds. We all loved him, as he'd keep us entertained with his Louisiana humor, no matter how stressed we got.

Our team's cadaver was a large, muscular man who required a lot of strength to dissect. Some cadavers were little old ladies who were so thin it was pretty easy to work on them. But this guy took a real team effort.

When we opened up his skull, we got quite a surprise. He had a large fluid-filled cyst about the size of a tennis ball in his left frontal lobe. Toward the end of the term, we all learned a bit of background on our cadaver. He happened to be a physician who had committed suicide.

That cyst had no doubt led to his suicide. Cysts like that in the frontal lobe of the brain are associated with dramatic changes in mood and behavior. I thought we ought to relay our findings to the family right away so that they would know that the man they loved had taken his life not because of anything they had done, or because of some terrible secret he was keeping, but because of an undiagnosed medical condition. The cyst in his brain had made him kill himself.

When I told my anatomy instructor that I thought we owed it to the family to tell them of this finding, he told me it was against the agreement the medical school made with all people who donated their bodies—namely, that under no circumstances would any findings from the dissection be revealed to anyone on the outside.

"But that's to protect them from findings they would not want to know about," I protested. "This is a finding that would very likely help them to make sense of this man's suicide and give them some peace of mind. We owe it to the family to let them know."

My instructor was a haughty surgeon who also had a Ph.D. in anatomy. He held himself in the highest regard. "If you reveal anything about this man to anyone outside the medical school, I will personally see to it that you are expelled."

"But that doesn't make sense. This information can only help this man's family."

"You heard what I said, and I mean it," the instructor said, and strode away, his authority established.

I didn't think it was worth getting kicked out of medical school to inform the family of what we'd found, but I always thought we should have done so. The comment my instructor wrote for my record read, "This student was more interested in the social history of the cadaver than in the anatomical details." Although I passed Anatomy, he was right.

43.

Although we had little exposure to patients in the first two years, in the second year I was able to take an elective in psychiatry and take on the first patient I ever had.

I was assigned a forty-year-old man with obsessive compulsive disorder. I was just learning what that disorder was, reading a couple of texts as well as MacKinnon and Michels's classic, *The Psychiatric Interview in Clinical Practice*.

The patient, Hank, needed more than my knowledge from a text. He was crippled by his condition. He could barely leave his house. It's not uncommon in severe OCD for people to be such slaves to their compulsions that they can't even go outdoors, which is another disorder called agoraphobia. Hank had that, too.

He was deeply embarrassed by his compulsions, but he found them impossible to resist. His only means of coping became staying indoors, at home. He'd lost his job as an insurance adjuster. His wife was getting fed up, and his three children were starting to mock him.

He told me his story as we sat in a tiny office, barely big enough for the two of us, that was reserved for medical students and their patients. The office had no window, just a desk and two straight chairs, as well as a telephone.

I was able to read some of Hank's history before I met him, but not the details of his compulsions. I just knew he'd sought treatment for OCD for a year, and nothing had helped. I was surprised, given all he'd been through, at how friendly Hank was. About five foot ten, with a wiry build, he wore steel-rimmed glasses, a V-neck sweater, and loafers. Until he got into the details of his problem, you'd have had no idea there was anything unusual about him at all.

But then, he got into the specifics. "Doc," he said, his hands clenching as he proceeded, "I can't walk past an ashtray without stopping, picking it up, and licking it. If it's one of these ones that's secured to the floor, I have to lean down and lick it anyway. Ashtrays are everywhere, so I can't go anywhere. And then my other compulsion is when I take off my underpants, I have to put my face into them and smell them. This one isn't so bad because I do it in private, but it's still disgusting and I hate myself for doing it."

I wanted to burst out laughing. *You've gotta be kidding me*, was what I was thinking inside. Thank God my childhood training in politeness kept me in check. As it turns out, *You've gotta be kidding me* is pretty much every sane person's honest response when they hear the story of a person who has a serious mental illness.

But as he continued to talk, and I saw how ashamed he was and how constricted his life had become, my initial naïve reaction changed to empathy and concern. Hank was really suffering. I also was worried because I had no idea how I could possibly help him. If all the pain and embarrassment he suffered was not enough to get him to stop obeying his compulsions, what on earth could I offer?

"You can imagine," he said, "how totally embarrassed I feel when I feel I have to pick up an ashtray and lick it, even with someone looking right at me when I do it. Some of them say right out, 'That's disgusting!' and I say, 'I agree,' as I put the ashtray down and walk on. Thankfully, with the underwear, no one is looking, but I feel totally ashamed and disgusted anyway. Why do I do this, Doc? I'm really sick, aren't I? Can you help me?"

Other experienced psychiatrists had tried and failed to help Hank. He could no longer afford insurance or a private fee, which is why he was seeking help through Tulane and Charity. To top it all off, now he was stuck with me, a second-year med student who didn't have a clue.

But I did have backup, thank God. In addition to a resident who supervised me, I could also turn to a senior member of the faculty, Dr. David Melke. When I presented the case to Dr. M., he asked me what I wanted to do. "What I *want* to do is tell him to stop it," I said. "But I know that's stupid. No one wants to stop these habits more than he does."

Dr. M. responded, "In years gone by, this man would have been put into psychotherapy, maybe even psychoanalysis, because that's all there was. It wouldn't have worked. In really bad cases like this one, people sometimes committed suicide."

"Why does he do this?" I asked. "Is there *any* way we can help him?"

"Why he does it we can't really say, except that it is part of a well-established diagnostic entity called obsessive compulsive disorder. You should read about the difference between the obsessive compulsive style, which MacKinnon and Michels write about in the book you're reading, and obsessive compulsive disorder, which is what's crippling this man."

"Insight won't help much, will it?" I asked, knowing that it wouldn't.

"Probably not very much, but your support definitely will. He needs to be able to talk with you and know you are not going to laugh at him. Everyone else, his friends, his family, the world, they all ridicule him. He ridicules himself in all likelihood and finds himself disgusting. You need to help him understand that what he's got is caused by biology, not lack of willpower or hidden unconscious drives. You can help him a lot by explaining to him that what he suffers from is more like a seizure disorder than anything else. He simply can't control the compulsions any more than a person can control a seizure."

"OK," I said, "I can do that. I can offer him support, which I am actually doing, even though my first reaction was to laugh at him inside."

"You can learn from that reaction," Dr. M. said. "It's the reaction he brings out in most people, so knowing that, you can understand him and support him even better."

"But is there anything I can do beyond that? Other experienced doctors have tried. I feel kind of out of my depth here."

"Welcome to psychiatry," Dr. M. said. "We all feel that way often, because we don't have the cures we need for lots of the problems we help people deal with. But in this case, I think there is something you could try that hasn't been tried yet."

"Really?" I said, feeling a surge of hope and excitement. "What?"

"The MAO inhibitors have been found to be useful in treating OCD, and I don't believe he's ever had a trial of an MAOI, has he? If that doesn't work, we might consider psychosurgery."

"Surgery? Really? Well, in any case, he hasn't had an MAOI yet," I said. "Let's give that a try first." Dr. M. agreed. So began my first excursion into the world of prescribing. I couldn't sign the prescriptions of course, the resident did that, but I could do everything else. Like worry about side effects.

Monoamine oxidase inhibitors, or MAOIs, can help OCD, but they have an unusual, and to me totally nerve-racking, side effect profile. If the patient taking an MAOI eats any food with the amino acid derivative tyramine in it, his blood pressure can shoot through the roof and he could have a stroke and even die. It's called a hypertensive crisis, and it's potentially lethal.

So I had to go over in great detail the list of foods Hank had to avoid, lest he possibly die. This was a tough conversation for me, but Hank was so eager to get some relief, he didn't mind it one bit. No fava beans, OK. No aged cheeses, OK. No red wine, OK. No smoked meats or fish, OK. No food restriction would cramp Hank's style anywhere near as much as the OCD.

Hank and I made a plan that he would be very careful with what he ate and he would check in with me on the phone every day to report on his progress.

At first I spent each day and night expecting to get paged that Hank was in the ER, his blood pressure through the roof! But that page never came. Instead, Hank phoned in or sometimes came in for our brief daily check-ins and started to report progress. If I was overjoyed, which I was, imagine how Hank felt. He cried. He hugged me. He reported making forays into the outside world and actually walking past ashtrays without even touching them, let alone licking them.

The MAOI didn't cure Hank completely, any more than our daily chats did, but they did free him up enough to go outside and begin learning how to control his compulsions.

People who disparage medications in psychiatry ought to meet the Hanks of this world, before and after.

I'll never forget him, my first patient-teacher in psychiatry. It also makes me smile, all these years later, how anxious I was having just *one* patient under my care (and with backup, no less) on a medication with one potentially dire, i.e., lethal, side effect. We do begin with baby steps.

It's every pre-med student's fear and every med student's fear, and on some days, every experienced doctor's fear: How do you handle all that responsibility? If you're a car mechanic and mess up, or if you're an attorney and mess up, it's unlikely someone will die.

But if you're a doctor? We all play it out in our minds until, at some point, we get used to it. It becomes ambient noise; we become skilled in living with the possibility of bad outcomes.

What I learned—starting with Hank, and then from the thousands of patients since—is that if you hang in there, you're handling it. Lyn's husband, Tom Bliss, the orthopedic surgeon and my earliest doctor role model, told me, "Patients are better off having you than not having you. Just don't walk away. And always ask for help when you need it. The more cases you handle, the more skilled and confident you become."

But even as a raw med student or intern, with so little experience, I never felt terrified by the life-or-death decisions on my shoulders because I knew I was never alone. I never had more responsibility than I was able to take on. If something came up that I couldn't handle, there was always somewhere to turn, someone to ask. If I'd been alone in the middle of nowhere it might have been different, but I never was, thank God. A few years later, one of my best teachers in psychiatry, Dr. Tom Gutheil, summed it up in one of the most valuable maxims I've ever learned: "Never worry alone," he'd tell us, "never worry alone."

44.

When the third year finally came, I started to love med school.

My first rotation was Surgery. As a third-year med student, I was paired with a senior resident named McCormick. He could not have been more different from the pompous surgeon who'd been my Anatomy instructor. He was earthy, blunt, and interested in helping people. No one called him anything but McCormick. He had a first name, but no one ever used it. I followed McCormick around for the three-month rotation and was at his beck and call for just about anything.

Although I was essentially his slave, I loved McCormick. I don't know why, because he routinely abused me. He always called me "dink" or "fuckin' dink," never Ned. He chain-smoked Marlboros, barked orders constantly, and asked questions I couldn't answer. But he also taught me a ton and I loved his out-there, no-holds-barred, take-me-or-leave-me honesty. While other residents were kissing up to faculty and angling for jobs, McCormick didn't appear to give a damn what other people thought. I loved that. It reminded me of many people in my family.

And, boy, did McCormick ever love being a surgeon. "I'd rather cut than fuck," he said to me more than a few times. "Surgery is so off-the-wall hard-ass and definitely gonna mess up your life. I mean, there is *no such thing* as a surgeon who has a happy normal life, you really gotta love it to do it. How many surgeons have happy marriages? I don't know any, do you, dink?" He was really asking me.

"I know one," I said. "My cousin. But he's an orthopedic surgeon."

"They don't count. It's easy to be happy if everyone you operate on gets better and no one dies. But how many general surgeons have happy marriages? Damn few, I can tell you that. And how many surgeons are obnoxious? Most of us. How many surgeons end up alcoholics, divorced a hundred times, alienated from their children, drowning in debt because they can't handle money, and have people everywhere they go pissed off at them? Most of us. But, lemme tell ya, we love what we do. We love to cut. Oh, shit, do we ever love to cut. We love to fix people. And most of all we love that we are better than all the other docs. All those wimps who don't dare do what we do. We try not to let it show, but I can guarantee you, deep within the heart of any surgeon worth his salt is a voice saying, *we rule*." Then McCormick would light a cigarette. "You get it, dink? We rule. Now go get Mrs. Lafitte's labs."

I had only three months with McCormick, but he left his mark on me forever. My best friend in medical school, Steve Bishop, who went on to become a psychiatrist as well, used to make fun of my love of McCormick. "Surgery is all sublimated sadomasochism," he'd say. "What you've got going with McCormick is a sadomasochistic relationship, pure and simple."

"Hey, dink," McCormick asked, smoking the ubiquitous cigarette while we were having a coffee outside Charity at the aluminum truck on Tulane Avenue, "have I turned you on to surgery yet?"

"You have done that, Dr. McCormick," I said. "You've definitely turned me on to surgery. Although the stench of that anaerobic abscess will follow me forever."

Another drag on the butt, another sip of coffee, then the butt drops to the sidewalk and McCormick steps on it with his penny loafer (no socks). "What I told you about how surgery fucks up your life? Guess what—all of medicine fucks up your life. Beats the hell out of me why so many people wanna be docs. They go in thinking they can save the world, or make a lot of money, or get respect, I dunno, but I can tell you one thing for sure, they have *no idea* what they are getting into. Most of 'em end up burned out, drugged out, divorced, alone, miserable, it's a shitty story what life has in store for most of us. So you damn well better *love* what you go into if you want to stand a fuckin' *chance* of beating the odds. That's for *damn* sure."

I wanted to give McCormick a hug, but I thought he'd probably knee me in the groin if I did that, so I just said "Thanks a lot." I don't know where he is today, forty plus years later, but I'd like him to know he did OK by this dink.

Like the world of McCormick, medicine at Charity was a unique world. Studying there afforded opportunities not only to meet gifted, gritty teachers like McCormick, but to work with patients in a much more responsible role than most medical students at other schools got to.

For example, on one steamy August night during my OB/GYN rotation (I did OB after Surgery with McCormick), the air conditioning in Charity shut down. New Orleans in August is like a tropical rain forest. Without AC, the hospital air was all but liquid. As the night wore on, I had to change my scrubs over and over because they were sopping wet from my sweating.

If I was hot and dripping, just imagine what the women in labor were going through. We all worked together to keep them as comfortable as possible, but the words "comfortable" and "labor" don't usually appear in the same sentence.

Around three in the morning, a woman in her thirties came in to deliver her fourteenth baby. She was assigned to me. I brought a cold washcloth to her bedside as I introduced myself. She gave me a big smile as she gladly wiped her face with the cool cloth. "Sweet Jesus, did the AC go out in this place? Oh, Lordy, I don't know what I'm gonna do, did it really go out?" she wailed.

"I'm afraid so," I said, and then, after looking down at the chart, added, "Harmony."

"Well, this baby ain't gonna wait, I can tell you that. I don't need you to do much, though. I've done this a few times before." With a loud grunt, she shifted her large frame around on the cramped gurney, trying to find the least awkward position. "There ain't no easy way, you just want it *quick*. And I don't want no drugs, and *no* 'pisiotomy. They just *love* 'pisiotomies here at Charity, and I'm telling you, *I don't want one! You hear me?*"

"Yes, ma'am, I do," I said as forcefully as I could. She was correct. The Tulane department of OB at the time mandated that all vaginal births include an incision diagonally and downward from the vagina, a procedure called an episiotomy. It was supposed to reduce the risk of vaginal tearing, which, if it does occur, can create a big mess. And with medical students doing most of the routine deliveries, the policy was intended to prevent complications that medical students like me couldn't handle.

But I believed this woman had had enough experience giving birth that I could obey her command and simply let the process go its natural way. If I got into trouble with the resident or the attending, I'd just tell the truth. The patient does have the right to refuse a procedure, after all.

All of a sudden, Harmony let out a whoop. "Honey, it's *coming!*" At that moment my hand was inside of her, using my fingers to measure how dilated her cervix was. "I'm plenty dilated up, if that's what you're checking, or maybe you just having a good time down there!" She let out a hoot/scream of laughter/pain and then gave my free hand a titanic squeeze.

Next thing I knew she was crowning and the baby shot out into my waiting hands.

"It's a boy!" I announced, trying to add something useful to the procedure.

"My twelfth boy. I'm a regular dingaling factory. Clean him up good now and let me hold him, you hear me?"

With the help of the nurse, we suctioned and patted and dried. Based on skin color, reflexes, muscle tone, heart rate, and quality of respiration, the Apgar score is assigned after one minute and then again after five. This baby scored 10 and 10. A healthy baby and a healthy mom, especially for someone who'd just delivered her fourteenth baby.

He was adorable. Actually, I found *all* the babies adorable. OB was by far my favorite rotation in medical school, even more than my months with McCormick. In that one month of August, I delivered thirty-five babies.

When I handed this squinchy-faced infant to his mom, Harmony looked down at him and heaved a sigh. Something was bothering her. As her eyes flooded with love—unmistakable, even with number fourteen—she looked up at me with a troubled look. "Doc, you could help me with one thing."

"Anything, just ask."

"Honey, I am fresh out of names. I come here expecting a girl, don't you see, and then I get another boy. He's beautiful for sure, but I just don't have a name for him. If he'd been a girl, she was gonna be Georgina, after my best girlfriend, by the name of Georgina, of course, but seeing as he's a boy, well, Georgina won't do, and I just plain do *not* like the name George. So I was thinking as I been laying here, maybe that nice doctor's got a name for my baby?"

In the middle of the night, in the sweltering heat and humidity of Charity Hospital in August with no AC, my brain was on tilt.

Without a second thought, without even pausing for a millisecond (years later, I would learn this is what we people who have ADD are prone to do), I proposed, "How about Fenway Park?"

"Ho!" Harmony crowed. "How about that! That's a right classy name. I love that. How you say, again, Fenway Park?"

"No, no," I sputtered, "I was just joking, you can't name him Fenway Park. That's the name of the baseball field in Boston, the city I come from up north."

"I know where Boston is, child, one of my sisters lives there with her fat, no good husband. Oh no, I take that back. Lord, forgive me. He is a *good* man. He treats her fine. Just some days I wish he could *work* harder instead of laying on his fat, no 'count butt. Shut your mouth, Harmony! Lord, forgive me. Anyways, Doc, I'm glad to name this boy after a place in Boston in honor of my mighty fine sister."

"Really, I don't think you want to do that. Fenway Park is the name of a *baseball field*. He's gonna be stuck with this name and he may get made fun of because of it and he may not thank you for giving it to him. In fact, he may be very angry with you, not to mention me! Please, I can come up with some normal names for you. Bob, Joe, Frank—"

"Too late, Doc. He can change his name if he wants to when he gets old enough. But for now, I am in charge and my new son's name is Fenway Park. It's a gift you've given me, Doc, it's a right classy name, and I am grateful to you and to the Good Lord for sending me this beautiful name for my beautiful son number twelve."

45.

ike my own family, Tulane Medical School, Charity Hospital, and the city of New Orleans teemed with wildflowers: colorful characters not traditionally cultivated. Not many days passed without a funeral procession dancing down one of the streets in the Quarter, mourners dressed in motley array, men in black top hats, women sporting parasols of many colors that they twirled and swung around as they sashayed along, accompanied by the brass instruments playing slow hymns like "Nearer My God to Thee," followed by upbeat tunes like "When the Saints Go Marching In."

I loved watching these processions from my dorm window, or even better, while sipping a beverage in one of the many raucous but welcoming bars scattered throughout the Quarter. My God, I thought to myself, *this* is the way I want to go. Why on earth don't we white folk do it like this? I could see people with tears streaming down their faces even while they were pirouetting and dancing. And isn't that *exactly* what a funeral should be? Tears over the loss, weeping that you will not see the person again, at least in this world, coupled with a celebration of the life the person lived, and jubilation over how much love all the people gathered felt and feel for each other and for the person who's passed away? At Exeter, the English Department had a blanket policy against using euphemisms for "die" like "pass," yet these New Orleans funerals seemed to defy the idea of dying and celebrate the idea of passing on.

The cuisine in New Orleans was as distinctive as the funeral processions, and just as much of a celebration. It has evolved since I was there in the late 1970s, influenced by the genius of chefs such as

Paul Prudhomme, Emeril Lagasse, and "Miss Ella" Brennan, but even when I lived there it was unique in the world of food and truly to die—or pass on—for. The lowliest bar served a great red beans and rice, every restaurant served coffee with chicory so rich you could almost use it as a sauce, the baguettes were crunchy, warm, and chewy, and the service was always begging us busy, serious people, "Just sit back, relax, and *ennn*-joy. Don't hurry none, y'hear?"

I had my favorites. When I could afford them, the famous Galatoire's with its perfect shrimp rémoulade; Commander's Palace with its Cajun-spiced pan-seared redfish; the Caribbean Room at the Hotel Pontchartrain with its inimitable Mile High Pie; Pascal's Manale with its luxurious barbecued shrimp drenched in a bath of olive oil, white wine, garlic, and Manale spice (black pepper, cayenne, paprika, salt, thyme, oregano, and basil); and all the fish shacks on stilts out on the lakeside—all destroyed in Hurricane Katrina but since rebuilt.

And then there was a very special joint—it looked like a joint for sure, a ramshackle shack on the side of the road which you'd never know was one of the best restaurants in the USA—that you had to drive the twenty plus miles across the causeway to get to, Mosca's, with its fabulous, nowhere-else-to-be-found Oysters Mosca as well as its simple crabmeat salad that is like no other crabmeat salad you will ever have and that you will never be able to replicate. (Believe me, I've tried dozens of times.)

Like New Orleans' cuisine, many members of the Tulane Medical School faculty were unique. Aside from my favorite, McCormick, there was Thorpe Ray, chief of Medicine, whom students called Yoda. He was a barrel-chested Texan who smoked cigarettes (you had to be one stubborn, addicted, independent-minded sumbitch to be chief of medicine at an academic medical center and smoke cigarettes in 1978) and could diagnose just about anything within minutes of hearing the history. He could gauge your degree of heart failure by looking at your neck veins from across the room, could tell your hematocrit within a point by looking at your nailbeds.

He had no use for the abstract. One time I went with a few other students to talk about the moral questions raised by keeping people alive when they wanted to die. He listened to us, quietly smoking his Camel, then put it out in the ashtray on his desk. "You're third-year students doing your Medicine rotation, am I right?"

"Yes, sir," we said.

"Well, my friends, here's my advice. I suggest you spend your days and burn the midnight oil learning how to keep people alive. When you master that, then you can come back and we can talk about how and when to let them die."

Senior to Thorpe Ray was Robert (Bob) Heath, a movie-star-handsome, tall, tanned, white-haired psychiatrist who became a psychoanalyst because that's what you did in the 1940s but then became a neurophysiologist. He believed that mental illness had its roots in biology and could best be helped through biologically based therapies. At first, this made him a renegade. He headed to Louisiana, no doubt because he wanted to find the freedom to pursue his interests in ways that the more orthodox northeast medical community would not allow. A persuasive leader, he founded the Department of Psychiatry and Neurology at Tulane in 1949.

Heath was a brilliant man, and, coupled with his striking good looks, he exuded charisma. He could get away with just about anything. But, like many charismatic, brilliant people, sometimes he went too far and did what today we would look upon as unethical and deplorable. For example, he once tried to convert a gay man to being heterosexual by surgically implanting electrodes in the pleasure center of his brain, then hiring a prostitute for him to have sex with. We now look with horror at what he did, but this was the era when homosexuality was still regarded as a mental illness. I say that in no way to excuse Heath—what he did was hideous—but to put his actions into historical perspective. The treatment of gay people back then was cruel and unforgivable, not just in the field of psychiatry but in ordinary life everywhere.

When my cousin Jamie was a junior at Harvard in 1970, wrestling with the sexual feelings he'd had for boys and men since he was a small child, he went to the leading psychiatrist at the Harvard Health Services who after just a few minutes told him not to be silly, he was not homosexual at all, and to put those fears to rest. It's that kind of advice that precipitated suicides. It would be decades before the field of mental health would de-pathologize homosexuality and see it as what it is, a normal and healthy variant, a trait some people are born with.

Heath was trying to understand all of human behavior and emotion from a biological perspective, and in that regard he was ahead of his time. It's just that his methods and his findings were objectionable. In the 1950s he isolated an antibody he named taraxein in the blood of some schizophrenics that produced schizophrenia-like symptoms when injected into monkeys. This led Heath to postulate that schizophrenia was an autoimmune disease.

When he presented this idea to us med students in a lecture, my classmate, Tom Garland, one of the brightest guys in our class, asked, "If it is an autoimmune disease, wouldn't you predict that steroids would help, seeing as how steroids help most autoimmune conditions?"

Dr. Heath had to admit that that was a good point. There was, and is, no research to show that schizophrenics benefit from a course of steroids. Since then the role of taraxein in causing schizophrenia has been debunked, with many experts not believing that taraxein exists at all.

Heath later teamed up with a neurosurgeon to implant electrodes in the limbic area of some schizophrenics' brains. The limbic region controls emotion, and Heath's hypothesis was that there is some lesion in the pleasure circuitry that impairs schizophrenics. After surgery, the patient could push a button on the external box he or she carried around and stimulate the limbic system when the patient was hallucinating or otherwise having symptoms. I actually had one of these patients under my care as a student during my third-year psychiatry rotation. The patient was proud of the little box he carried in his shirt pocket, but beyond that, he reaped no benefit from his electrical stimulator. As far as I could see, the research subjected the patient to considerable risk with no gain whatsoever.

As a med student, I didn't know what to make of Dr. Heath. On the one hand, he was chair of the department of the specialty I would likely enter. On the other hand, he was going off on wacky tangents supported mostly by his own grandiose vision of himself. Or so it seemed to me. I found myself wishing, once again, that I had a less crazy, wiser man in charge of the family I was planning to join.

But Heath was the exception. Almost all of the teachers I had at Tulane were smart, ethical, and inspiring. They taught me doctoring, they taught it in a way that made me love doing it. I adored George Bailey, a professor of medicine who did his fellowship in nephrology at the Brigham in Boston before coming to Tulane. Along with three other fourth-year students, I did a clerkship with him senior year.

George was larger than life, all three hundred pounds of him. He loved food and wine. He was a true Dionysian, and his enormous girth showed it. He was a life enthusiast, brimming with energy and loving to teach.

Four of us seniors spent a week with him on the road. It began with his holding his two little daughters in his arms outside his home on St. Charles Avenue, where we met him, kissing them goodbye, and then piling us into his capacious BMW. He ran dialysis units in New Orleans and throughout the state and made tons of money, so lots of people were jealous of him, but as far as I could see he was a great doctor who happened to be a crackerjack businessman as well.

And boy, did he love food. He'd find the one restaurant in the middle of bayou country that was a gem. He was the person who introduced me to Mosca's and several more out in the swamps that I fear I'll never get back to but hope I will.

That week, as we drove those Louisiana back roads, he'd talk about food and wine (he taught me about the great Sauternes from Château d'Yquem, which I had never heard of until then), or give us practical tips on medicine. "I'm gonna tell you what you need to know as a doc but they don't teach you in your med school classes. Let's start with constipation. You've gotta have plenty of remedies for constipation in your mental Rolodex. People, especially old people, *especially* old people in rural Looosiana, don't like to be stopped up. And if you can give 'em some relief, they will love you forever. There's softeners, there's lubricants, there's stimulants, there's osmotic agents, there's bulking agents, there's enemas, and don't forget, there is always your index finger. Never forget your index finger. You can make an old person practically have an orgasm if you disimpact them right there on the spot."

"Aw, c'mon, George, I don't wanna be doin' that!" one of us said.

"I'm jus' tellin' you, you can bring a body a ton of relief, but you do what you want. And then we got headaches. You gotta have remedies for headaches. The two most common reasons a person goes to the doctor are cough and headache. So you wanna have good solutions for both of those."

I had more fun with George Bailey, and learned more that week about practical medicine, than during just about any other week of med school.

But in remembering the Wild West style of Charity and Tulane, I have to end with Dr. Gallant. Raised in Brooklyn, Dr. Donald Gallant came from New York City to Tulane to practice psychiatry and teach. Another connoisseur of fine food, Dr. Gallant even had a dish at Mosca's named after him.

I had only one direct exposure to him, but it's one I will never forget. I took the elective he offered in the treatment of addictions in general and alcoholism in particular.

Gallant was an immensely popular instructor, winning the best teacher award many times. He gave his all to us medical students and to the residents. Innovative, irreverent, but also ethical and conscientious, Gallant was a brilliant teacher and hugely generous, good man.

His theoretical basis in treating addiction was to start by breaking down the alcoholic's denial in order to help him or her achieve and maintain sobriety. This is standard procedure in the field. How he did it, however, was anything but standard.

It was theater. All of us—Gallant's sessions usually had an audience of around twenty-five, including students from all disciplines, interns, residents, and an occasional faculty member—gathered in a conference room and sat in a semicircle. Gallant sat in front of us, at the center of the semicircle.

At the start of the first session I attended, after Dr. Gallant took his place, a man entered the room along with his wife. The two of them looked to be in their early forties, maybe; both were attractive, in pretty good shape, and looked well off. The man was wearing a wrinkled suit and a nondescript tie, while his wife wore white pants and a pale blue blouse. Certainly not your stereotypical alcoholic and spouse. They did both look nervous, understandably, before this audience of strangers. Of course, they had been told what the setting would be, and they had agreed to it.

After pleasantries, the session began. "Have you had anything to drink since our last meeting?" Dr. Gallant asked.

"Well, I had one slip," the man sheepishly admitted.

Gallant then went off. "What do you mean, a slip? Did you slip on a banana peel? Or a cake of soap? Or did someone pry open your mouth and force alcohol down your throat? Or, just maybe, are you telling me you were so weak you just couldn't stop yourself?" He spoke those last few words in a mocking, whining child's voice. "And so you just *had* to take a drink? Aren't you ashamed of yourself?"

The man nodded.

"Well, you should be!" Gallant went on.

Then he turned to the wife. "Have you been giving him his Antabuse?" She nodded. "Did you give it to him that day?"

"I thought I did," she said. "I mean I thought I saw him swallow it."

Gallant went off again. "How many times do I have to tell you? When will you learn? Addicts are liars! They are born liars, they will do *anything* to deceive you." Then, looking at the man, "Am I right? Aren't you a liar? Won't you do anything to get your precious drink? Well, c'mon, won't you?"

The man nodded, almost imperceptibly.

"Well," Gallant went on, "what are we supposed to do with you? Are you gonna string this lady along forever, just lying your way into a divorce and losing everything?"

"No," the man said. "I'm going to do better."

"That's what you all say," Gallant barked. "Why should anyone here believe you?"

It was painful to watch this. I wanted to stand up and say "Stop it!" but restrained myself, knowing that I would only get into a ton of trouble and Don Gallant would only redouble his efforts to "break down the denial." I had to remind myself that Dr. Gallant had developed his method over years, and that people came to see him because he got good results.

I've thought of Don Gallant often in the years since I graduated. Was he right or wrong to use his confrontational style? I didn't know then, and I don't know now. I do know that I admired Dr. Gallant because he was doing his best to address the deeply difficult problem of alcoholism.

I am sure he did not think he was being cruel. He thought he was doing what needed to be done. That I felt it might have been cruel, well, that's just me. And Gallant had a whole lot more experience in treating alcoholism than I did, that's for sure.

But what he did teach me, and why I remain to this day grateful to him, is that the confrontational style with which he treated that patient, and all the others he ran through his regimen, lives within me just as much as it did within Dr. Gallant.

Because whenever he was going off on a patient, attacking and even humiliating him in front of all of us, carrying on as he did, all I had to do was imagine Uncle Unger in the patient's chair, and then, presto!—I'd be cheering Dr. Gallant on as if he were my knight in shining armor. "You tell 'im, Don, you sock it to him good, you bring him to his knees and make him beg for mercy. Don't let him worm his way out. Make him confess!"

Dr. Gallant showed me far more dramatically than any textbook or classroom lecture ever could have the rage that I felt toward Uncle Unger. Gallant got me out of my head and into my emotions. He made me see just how primitive a person I can be. Gallant was using his own aggression constructively, but I doubt I could have done that. I would have unleashed an unholy torrent of anger had I let myself go. Gallant was more disciplined than I could have been.

I took from those sessions with Dr. Gallant not just that I have the capacity to be cruel, indeed the desire to be cruel, but that I had better watch out for it. I had better learn to control it so it does not control me.

So while I cheered Dr. Gallant on, seeing Uncle Unger in the patient's chair instead of the patient, I came to learn that what I was cheering for is a part of me I should get to know and work to control rather than give vent to in the name of delivering care.

46.

In the winter of 1978, my fourth year of medical school, I got a letter from my father telling me he had lung cancer. "It was only diagnosed a few weeks ago," he wrote. "It started with a cough that wouldn't go away. Then they found a tumor in the upper lobe of my left lung. They say it's highly malignant. I go for chemotherapy twice a week. I stop the car, open the door, and puke every time on the way home. They say there's no point in doing surgery."

So he was dying, I said to myself. I went off to the public library just a couple of blocks from the medical school, sat in a chair, and stared out the window. The library was all glass windows. It was one of my favorite haunts, and I wanted to be alone.

He was always good to you. He saw you as much as he could. He took you to Red Sox games, he taught you to fish and to sail, he taught you to skate. If you'd had more time with him, he would have turned you into a good hockey player, maybe even as good as he was, not likely, but maybe. He played miniature golf with you, he took you and Jamie to horror movies in Hyannis, he played croquet with you and let you win. He was as good a dad as he could have been.

Then why did he let the whole thing with Unger happen? Why didn't he beat the crap out of that son of a bitch? Why didn't he come down to Charleston and take me out of there?

Because your mother wouldn't have let him. And because he had a major mental illness. It was pretty damn impressive he could even hold his job teaching school.

Why didn't he beat up Unger? He left defending yourself up to you. After Mom went off with Unger, Dad checked out. But he didn't completely; he sent her a dozen pink roses on her birthday every year.

It was all fucked up, Ned. You gotta see that. You were dealing with crazy people. Normal people try to get custody if their kid is in a bad situation, but these were not normal people. Gammy Hallowell had the money to send you to places where other people would look out for you, and you should thank God for that.

There was no bad guy, the badness came from madness. Dad was crazy, Unger was crazy, Mom became crazy, the whole scene was just plain nuts.

So now Dad's dying. At that moment I remembered Dad standing on the periphery of the soccer field at Fessenden, a hundred yards away from me, in the mist, staring out at our practice, and then, after I looked away for an instant, disappearing. That's the best he had. You should be glad he had that.

I should go say goodbye. I told the resident in charge of rotation that I needed a few days off because my father was dying. He said sorry to hear that, take all the time you need. This was fourth year of med school, no pressure.

I flew to Boston, rented a car, drove up to the hospital in New Hampshire. At the information desk, I asked the volunteer where Mr. Hallowell's room was. It seems as if all of that happened in about ten minutes.

The next image is me walking into Dad's hospital room. There he was, sipping something out of a paper cup through a straw, the upper half of the bed elevated, propping him up. He didn't look almost dead.

The moment he saw me he put his cup down on the bedside table and beamed. In as strong a voice as he could muster, trying his best to sound like his usual self, he said, "Oh, *John*, thanks *so much* for coming!"

Dad mistook me for my brother. Missing only half a beat, I jumped right in and said, "Hi, Dad," as I gave him an awkward, bedridden hug.

"It's so *good* to see you, John. It's been too long."

The fact is that he hadn't seen Johnny in years because he couldn't accept Johnny's being gay, nor could Johnny accept Dad's difficulty with it. If either one of them had made an effort, I'm pretty sure they could have made peace in a matter of minutes. But they hadn't.

My father's mind played a lovely trick on him at the end, like a mental white lie. Maybe influenced by the medications he was on, he transformed me into Johnny. He was at peace with me, Ned, he knew I loved him and he loved me, so there was no unfinished business between us, but with his son John, there was a searing need to reconcile. With me filling in, Dad did it. I'm just glad I was quick enough not to correct him.

In the car, driving back to the airport, I felt like crying but couldn't. I've never been a good crier. *It was good that you let him believe you were Johnny*, I said to myself. That way he can die feeling at peace with his three sons. Too bad he couldn't make peace with his brother.

The funeral was held at Mount Auburn Cemetery. I would later learn that most of the teachers at the public school in Pelham, New Hampshire, where he'd taught for many years, had wanted to come to the funeral. The principal called the district superintendent for permission but was told only four teachers could go because the school day had to go on, and it would be hard enough to find subs for four.

The teachers got together and went to the principal, saying that either the superintendent had to close the school and let all the teachers travel to Boston for the service or the entire faculty would go on strike.

Not only did the school close for the funeral, but the superintendent supplied two school buses to transport all the faculty and staff who wanted to attend. We were dumbfounded when two yellow school buses pulled onto the grounds of Mount Auburn Cemetery and dropped off a huge contingent from Dad's school.

"He was beloved," as assistant principal told me. "We never had a better teacher or a more cherished teacher. We really would have walked out if they hadn't let us come."

47.

All my life I've had a terrible habit of telling the truth as I see it, blurting it out, you could say, regardless of how inappropriate it might be. It's gotten me into more trouble than I care to recount.

"Why do you want to become a psychiatrist?" That was the question I was asked in every interview at every training program I applied to during my fourth year of medical school.

This was the time to say something like "Because I have always been interested in the workings of the mind," or "I had a great mentor in medical school who got me excited about psychiatry," or "The brain is the last frontier of human discovery and I want to join the exploration." All of those are solid, sane responses to the question. One of those, or one similar to those, is what I should have said, and what I would have said had I learned the practical skill of being politic and appropriate. But that skill never came easily to me. Being honest came so much more naturally.

So my answer to that question was "Because I come from a crazy family." Most interviewers stopped dead in their tracks, as if I'd just farted. Then they'd find some excuse to pause and regroup.

I remember one man in particular, who looked to be a junior faculty member, only a few years older than me, nattily dressed, wearing a bow tie. When I told him I came from a crazy family, he took off his eyeglasses and nervously started to clean them with the balled-up hanky that he pulled out of his trouser pocket as he tentatively asked, without looking at me directly, "Crazy?"

"Oh, yes, crazy," I replied emphatically. "They were definitely crazy. Lots of my type are. My family were all interesting people, just often drunk or crazy, or both."

This interviewer, having no clue as to what a fitting response might be, simply stared back at me. But telling him some of my story seemed important to me. I felt then what I've continued to feel my entire career, that psychiatry shies away from the real story and tries to dress it up in jargon and scientific-sounding nomenclature, as if to make it presentable to a general audience. Also, the point of an interview is for the program to get to know the actual candidate, not the tidied-up version, so I saw no need to tiptoe.

"My father was bipolar but first was misdiagnosed schizophrenic. He went crazy when he came back from World War II. He'd been an All-American hockey player at Harvard, led the nation in scoring, married my mother after a storybook romance straight out of *Love Story*, got a job at Goldman Sachs, had my two older brothers, but then went off to the war.

"When he came home, 'he wasn't right,' as they used to say. The fact is, he was nuts. So he was put in a hospital where he got insulin shock and electric shock. It was gruesome, but the doctors were just trying to help, that's what you did back then."

My bow-tied interviewer nodded, confirming a historical fact. I was glad at last to make some connection with him. But the next part of what I had to say really threw him for a loop. I told him the story of how I was conceived, complete with the dead crows and the summoning of the police.

That about finished him. He had clearly heard way more than he'd wanted, having twice taken off his glasses, wiped them, and put them back on while I was talking. I'm surprised he didn't break them in two, he polished them so aggressively. Now he was back to his default position, staring and at a loss.

"Of course, that was in Massachusetts," I said, "so you wouldn't know the hospitals where he was treated."

My interviewer said nothing.

Feeling the acute and throbbing awkwardness between us, which threatened to take the interview clear into the Twilight Zone, I tried to help out. "I'm sorry. I was just trying to be honest. You know, these interviews get pretty boilerplate, so I was just trying to be a little bit real. This is my third one today. I don't know. I hope it's OK. I probably shouldn't have said so much. But it is psychiatry, after all. But anyway, I'm sorry. I can understand if this all feels like too much. I have a habit of going overboard."

The man heaved a sigh, and his whole being seemed to defervesce, as if a mental abscess had just been drained. I could feel his relief as if it were a hug. "You're right!" he said. "This is my third interview, too. They are *so* boilerplate. You have no idea how tedious they can become. So thank you for changing it up a bit. Yes, you did that for sure." He chuckled, having regained control of the situation.

The tension broken, we ended up having a great interview, talking no more about my crazy father or crazy anybody, but safe topics like theories of the mind and the Yankees versus Red Sox.

Telling the truth—or your own version of the truth—is a bad habit if you want to get ahead in this world, but it's also a habit that—in spite of many punishments—I've not been able to shake. I wanted to become a psychiatrist because I wanted to understand my people in particular, and crazy people in general. I felt totally at home with crazy people, but I knew the world at large didn't understand them at all and felt anything but at home with them. I wanted to do something about that.

Based solely on statistics, the interviewer—any interviewer—should have rejected me from the program. The Adverse Childhood

Experiences study provides a score based on the number of specific adverse events in a person's childhood—like parental divorce, alcoholism in the family, mental illness in the immediate family, and seven others. A score of 4 or higher strongly predicts terrible outcomes in adulthood, like major depression, chronic unemployment, inability to stay in relationships, alcoholism, early death, and suicide. My score is 8.

Fortunately, statistics does allow for outliers. A few people—like me—beat the odds. I was accepted into that residency program and many others. Thankfully, the people in charge of these training programs paid more attention to my academic record and what I was like as a person now than they did to what the statistics based on my childhood would predict would happen to me as an adult.

Most people with my background who do survive do not thrive. They become cynical and pretty weather-beaten. They tend not to trust, to put it mildly. I, however, became a gullible, wide-eyed optimist. The price I paid is that I carry a lot of sadness inside me. But that also gives me a deeper understanding of other people's sadness that lectures and books can't provide.

Without knowing exactly why, I heeded the inner voice that persisted even after initially getting rejected from medical school, the inner voice that had visited the little boy that hot day in Chatham. My career path wasn't so much a choice as it was a necessity, a tropism, like a sunflower turning toward the sun.

48.

t was gloriously sunny in New Orleans the day of my medical school graduation in June 1978. Thank God the ceremony wasn't outdoors, as the weather was, of course, hot and humid, and we graduating seniors all wore black robes and caps with green tassels.

I was chosen by my classmates to give the student address at the ceremony, so I sat on the stage. The audience was packed with proud parents, grandparents, siblings, godparents, and other supporters. No one in my family was able to come, but that was OK, I would see them soon enough.

Graduation day at Tulane Medical School was a victorious day for me and for us all. We shared the can-you-believe-it feeling of *wow*, I'm actually a doctor now. Of course, we had a lot of work to do to live up to that title, but at least now we were in position to take on the task.

The transition from being a medical student to being an intern was intense. The first day of my internship I went into the men's room at the hospital, looked in the mirror, and said to my reflection, "Tell me why I shouldn't quit right now." My reflection said nothing back. I splashed cold water on my face and said out loud, "People dumber than you have done this, you can do this." So off I went, out onto the wards, hoping and praying I could pull my weight. The year turned out to be one of the best years of my life.

Of course, people died during my internship. People die in hospitals every day. For such a big deal, it's no big deal. One day I was talking with Lyn about it.

"Why is it like this?" I asked her.

"I don't know," she said.

"People die, and it's like just another event in the day," I said.

"But remember," Lyn replied, "you don't know them well, at least most of them. It can't matter to you as much as it does to their families or their friends."

"I know," I said. "It's just that I never thought dying would become such a routine event."

"It's all a matter of context," Lyn said. "You're working in a hospital. In a hospital, dying is pretty much an everyday occurrence."

"I know. I'm being stupid. I just can't get used to it." Then I paused. "Do you believe in heaven?" I asked.

"Heaven is bullshit," Lyn replied without a moment's hesitation.

"Oh, c'mon, no one knows for sure."

"Well, I'm pretty sure, Toots."

"But you don't know for sure. No one knows. I don't know. Don't you wish God would just make a big announcement so we all could rest easy? It would make life in this hospital a lot easier, I can tell you that."

"That would be nice," Lyn replied, "but didn't God supposedly already do that? There's this book that tells all about it, it's called the Bible, as I recall. Marnie called it the world's longest fairy tale."

"I don't know," I said. "I wish I did. How about if we make a plan? Whichever one of us croaks first"—that was Marnie's word, "croak"—"if they get to heaven, they have to do all they can to contact the other person, OK? So if I die, I have to send you some sign and let you know the good news, and vice versa."

"OK," Lyn said. "That's an interesting idea. But if I die first, it makes me sad to think of you waiting and waiting for a sign that will never come. Marnie was right. It's blackness, blackness."

"I know what Marnie said. So what's a good sign for us to use?"

Lyn paused. "Your mother always liked the British saying 'Pecker up!' How about if whoever dies first somehow sends the message 'Pecker up!' to the other person?"

"Pecker up. That's perfect," I said. "No way that could be mistaken for anything else. If I die, I send you a pecker up message, and if you die you send me a pecker up message."

"Sure. But as I told you, this is an amusing little deal we've made, but that's all it is, an amusing little deal. No one is going to be sending or receiving messages from the Great Beyond, I can guarantee you that. I don't know why you believe this nonsense. It's a crutch is all it is."

"I'm not sure I believe it," I said. "I'd just like to know. Let's just hope we have a long time to wait, that's all."

"Don't worry about that," Lyn said. "We do."

My whole family thought my musings about God were a waste of time. Jamie was the most avowedly atheist, while the rest were more agnostic. God was not a topic that came up much.

But I did turn it over in my mind quite a bit. I saw more death as an intern than in any other year of my life. Death almost became routine, but not quite. I'd always look down at the dead person and half expect him or her to wake up. And I'd want to say something, like "Good luck," or "Bon voyage," but I never did. I'd think those words, though, as I was disconnecting the lines.

Going from alive to dead can be sudden and dramatic if you get shot in the street or collapse of a heart attack while jogging, but most of the deaths I saw in the hospital were rather serene. The patient would often be alone. The moment of death only became dramatic when we decided to do an all-out code and use the full tympani section to keep the patient alive.

But when we let nature take its course and let the patient quietly slip away, the moment that separates life from death is as imperceptible as the moment when love, once and for all, finally ends.

49.

Remember that last day of my internship at the West Roxbury VA Hospital that started this book? I made rounds on that breezy sunny Saturday morning, wrote up my notes, and said goodbye to the nurses and my favorite patients, especially the slowly dying Mr. Cavanaugh. He had end-stage everything but was fighting his last battle with all the cussing and flirting with nurses he could muster. I said goodbye to the medical interns and residents who'd been on call with me, people I treasure to this day, and walked out of the VA hospital for the last time, the most intense year of my life complete. *Finito.*

I felt sad to be leaving behind the exactitude of internal medicine. But my calling was psych. I now had less than twenty-four hours to get ready for the next shift, which would last the rest of my life.

That Sunday, July 1, 1979, I would report to the hospital that I would embrace with a passion I'd never felt for any institution (except maybe the Boston Red Sox). Innocuously called the Massachusetts Mental Health Center, or MMHC, it was built in 1912—ironically, the same year as Fenway Park. Originally called by a more colorful name, the Boston Psychopathic Hospital, or "the Psycho" for short, it was renamed the blander MMHC in 1967 in the interest of reflecting a more 1960s-ish community-based mission. But to the old grads, which include some of the most illustrious psychiatrists throughout the country, it will always be "the Psycho."

A shrink is a hodgepodge of a job at the bottom of medicine's totem pole, a job most people, often for good reason, make jokes about or flat-out scorn. Yet for me it was the Big Leagues, Broadway, the Show. People told me, don't romanticize it too much because you're bound to be disappointed. But I ignored that advice. I romanticized the hell out of it.

Signing on at MMHC was the logical continuation of talking with Uncle Jimmy about why he went into farming instead of finance, and why he hated hypocrites so much; it was the logical continuation of fervent discussions in English classes around Exeter's round tables, the logical continuation of the unexpected excitement I felt while listening to Walter Jackson Bate bring Samuel Johnson to life, the continuation of talking about people around the dining room table, sizing them up, commenting on their oddities and charms, which is what my family liked to do more than anything else. We were a bunch of gossips and critics of the petty ways of people, including our own, as long as the person being discussed was not in the room.

Entering psychiatry, I dreamed of nothing less than decoding human nature. Just about everyone who came to MMHC dreamed that the place would cure whatever ailed them emotionally and reveal the answers to their most fervent questions about life. I was no exception. I dreamed of a world where, at the feet of great teachers, I would learn about what the rest of medicine couldn't have time for, what the interns I trained with and all the others in mainstream hospitals simply could not (and should not) squeeze into their day, namely how to delve deeper into another person's mind, another person's story, with an eye toward "curing them up," as Dr. Bill Beuscher, my attending, or what we called superchief, used to say.

How often I'd imagined this, when I could start in earnest, a knight in training, to learn how to do battle with the demons of the mind, including my own. Like Stephen Dedalus, I was ready to forge in the smithy of my soul . . .

I harked back to a moment in Mr. Tremallo's class at Exeter. One morning Mr. T. suggested there was parody in Stephen Dedalus's words that come at the end of *Portrait of the Artist as a Young Man*: "Welcome, O life! I go to encounter for the millionth time the reality of experience and to forge in the smithy of my soul the uncreated conscience of my race."

"Do you think Joyce meant us to take that proclamation at face value?" Mr. T. asked the class, fourteen sixteen-year-old males sitting around a massive oaken table. "Or was he poking a bit of fun at Stephen?"

"How could he be poking fun?" I piped up. "How can it be parody? This is Stephen's big moment. It's like his battle cry. Stephen meant those words with all his heart!"

"Why?" Mr. T. asked, in typical Socratic fashion.

"He meant it with all his heart," I repeated. "I think he believed in his dream and now he was announcing it for all the world to hear. I don't believe there was a drop of irony or parody in it at all."

Mr. T. paused, then replied, "I agree with you that Stephen meant it. He meant it the way *you* would have meant it"—Mr. T. paused and smiled, gentle as always when he challenged us—"in a totally sincere, young man's way. And since you're a young man, and all of you are young men, it's easy to miss how embarrassingly, ridiculously self-serious those words ring to an older person's ear. But do you think perhaps the author, the very clever James Joyce, was employing some artifice and making a little bit of fun of Stephen's ultra-grandiose vision? Not in a mean way, of course, but in a sympathetic way, parodying the grand dreams of the young. From the older person's vantage point of knowing such dreams rarely come true?"

"How can you say that, Mr. Tremallo?" I countered, almost shocked. The rest of the class watched as Fred and I engaged. "Dreams don't come true? How can you be so cynical? Dreams do come true! Why are we here, getting all this education, if not to make dreams come true?"

"You tell me," Mr. T. said with his characteristically wry smile.

"You want me to say something like dreams are just an illusion and self-deception, and we're here to learn that lesson, like we just learned it reading *Gatsby*, but I'm not going to say that," I replied.

"Any more than Stephen Dedalus would!" Mr. T. said, slapping the table to drive home the idea. "That's the whole point, isn't it? Joyce has enough distance on his hero, who is himself after all, don't forget that fact, to appreciate the incredibly jejune nature of his statement."

"What does 'jejune' mean?" I asked.

"Look it up," Mr. T. said.

While I was discovering in the gigantic brown hardbound Webster's Third New International Dictionary that "jejune" meant "naïve, simplistic," another student, Julian Connelly, spoke up, helping me out. "So he wants us to like Stephen for being such a dreamer and thinking so big, but he also wants us to see that Stephen is still young and has a lot to learn."

"So what happens when you learn a lot?" I asked. "You give up on your dreams and become a cynical old man?"

"No," piped in Rob Shapiro, the smartest kid in the school, "you become wise."

"So what's the difference between being wise and being depressed?" asked my friend Paul Zevnik.

"If you're depressed, you don't think," chimed in Paul Goldenheim, who was headed for medical school one day. "You just feel. And you feel like crap. But if you're wise, you also *think*."

"What's the good of that?" asked another kid, Eli Robertson, both poet and super-jock. "What's so good about thinking? Thinking is overrated. It's all anyone does around here. At this place, emotions are forbidden; all you can do is think. Thinking and being smart; what's the good of it if you're not happy?"

Mr. T. now chimed back in. "So was Stephen *thinking* when he wrote those words?"

"No way," someone else said. "He was like on a date—dreaming big, trying to get lucky. He'd lost touch with reality."

I remember not wanting to agree with Fred Tremallo back then (or now), even though I knew a sophisticated reader would agree with him.

Just as in psychiatric training when I would be cautioned that we have few cures, only salves, I would bristle, just as I would when told not to cheerlead but rather allow the sadness to flow and despair to fill the room. When told early on that every trainee comes to Mass Mental seeking the Holy Grail of human nature but no one finds it, I would take umbrage. It was years before I gave up (sort of) what the seasoned pros in mental health call rescue fantasies, naïve dreams of cures, and instead reluctantly settled for what minimal good I could do.

I was not, at heart, a sophisticate, at Exeter, or now, on the doorstep of residency. I was too much of a poso for that. I was as foolishly bold and naïve as Stephen Dedalus. When it came to sophistication, I always kept in mind the words of Samuel Johnson, who described sophisticates as people who are "too refined ever to be pleased."

I wanted none of that. I was an all-in type of guy, a man of, in my friend Jon Galassi's words, "wretched excess," intemperate, full of enthusiasm and a tendency to overdo, masking pockets of loneliness, depression, pessimism, and darkness.

Of course, through all my grand dreams of saving others surged a selfish desire, the crazy wish that I could have saved my father, that I could have kept the storybook romance he had with my mother from ending, that the five of us could have lived happily ever after.

50.

The first baby I fell in love with was Lyn and Tom's first child, Mary Josselyn, aka Molly. Being in college when she was born and nowhere near a baby-making age, I still knew, just from watching Molly and seeing the joy Lyn and Tom took in her (despite all the work!), that someday I absolutely wanted to create a family. It became my most cherished goal: someday to create a family and give my kids the happy childhood I wasn't able to have.

Years later Molly enrolled in the pre-med program at Bryn Mawr, got into Brown Medical School, and became an E.R. doctor, which she is today.

When Molly finished her training, Lyn decided she wanted to throw a graduation party for her daughter and her fellow graduates. Since the gardens at the Blisses' beautiful country house would be in full bloom, they decided to have the party at home.

Although from all external appearances the Bliss family was as WASPy and preppy as could be, Lyn really hated being typecast. She did not want to be just another photo in *Town & Country*.

She set up the party just as one would expect—passed hors d'oeuvres prepared by a top Providence caterer, formally dressed servers and bar staff, full open bar, music, everything you could ask for, and we even got beautiful weather. All just as it would have been drawn up.

With one exception. One detail set this party apart from all others of its kind. One element imprinted this event with the mark of Lyn, of Josselyn Hallowell Bliss. One of the waiters was naked. Well, not entirely naked. He wore sneakers, and a red bow tie around his bare neck. Otherwise, he was as naked as the day he was born.

Since I grew up in boarding schools and was in male locker rooms just about every day, male nudity does not grab my attention. Penises do not stick in my mind after I've seen them. But *this* one did. I can see it now, hanging there in all its full, flaccid glory, not too big, not too small, just right, a living tribute to Lyn's absolute insistence on being herself, convention be damned.

Here was where politeness got trumped by something bigger. Lyn was indeed a stickler for manners, as all of her children could tell you. She'd sit at the head of the dining room table at family dinner with a knife in her hand, one kid on one side, one on the other. They took turns for who was forced to sit next to Mom. Should one of them make the mistake of putting an elbow on the table, *whack!* came the handle of the knife, inflicting serious, memorable pain. Thank-you letters after birthdays and Christmas were to be written within forty-eight hours, and read over by Mom; they had to include at least one interesting detail. Napkins went onto laps the minute you sat down, pleases and thank-yous became as automatic as breathing by age five, and handshakes were rehearsed until perfect. Lyn deeply believed in manners.

Hiring a naked waiter to staff your party is not exactly out of Emily Post. However, it *is* out of the book by which all of us Hallowells were raised, the book that scorns hypocrisy more than anything else, and that values being genuine most of all.

It was fun to watch the various reactions people had when the naked waiter approached. Most laughed, and, playing along with the joke, took from his tray whatever he was passing around. But some people scurried the other direction, refusing any sort of interaction. Since Lyn invited the families of the graduates, not just the graduates, there were many older people there, some of whom harrumphed about the "tastelessness" of this "display." But others openly marveled at how daring Lyn, Tom, and Molly were to do this, making a standard party unique and most certainly unforgettable.

I'm pretty sure Tom would not have chosen to do it this way, but he always let Lyn be Lyn (he really had no choice), and he did love her for who she was. Molly took more after her father as well, in that, as fiercely competitive as she was, she was naturally reserved. But she also loved her mother for who she was. As much as Lyn could be bossy, manipulative, moody, erratic, and at times just plain impossible, we not only loved her but also felt grateful that she took big chances and added so much pizzazz to our lives.

She was the glue that held the family together. Once Gammy Hallowell died, there was no one else to bring us all together, at just about gunpoint if need be, until Lyn assumed that role. Every family needs such a person if they want to stay close and actually care about each other.

51.

A month into internship I decided to take the plunge and get psychoanalyzed. This was 1978. Psychoanalysis was no longer a requirement for psychiatrists in training, as it was in the 1950s and 1960s, but neither was it considered the impractical and obsolescent step it would become by the turn of the millennium.

Ever since I first learned in vague terms what psychoanalysis was back at Exeter, I had wanted to "get analyzed," as people put it. It seemed like that mysterious process might hold the keys to the kingdom, the keys to unlocking the secrets of the psyche. Maybe it was just the twentieth century's version of phrenology, or the theory of the humours, but I was hooked. I wanted in, even though I knew nothing substantive about it except that it was time-consuming, expensive, and not guaranteed to do any good.

Then why in the world did I so want to do it? I had two compelling, unshakable reasons: desperation and curiosity. I felt compelled to do it, as if getting analyzed were my best, if not only, hope of gaining release from my insecurities, depressions, and doubts and becoming a happy, well-adjusted adult. Reason number two, I absolutely *had* to satisfy my curiosity as to what this mysterious process that gripped the imagination of the world's intelligentsia was all about.

A swarm of cynics couldn't have held me back, not that any tried. In fact, almost all of my friends and colleagues, including all the people who were headed into the practice of medicine, encouraged me and told me they wished they could do it as well. Most intelligent people want to know what makes them tick.

Since I was doing my internship through the VA system, and the VA insurance still covered psychoanalysis, my first eight months or so would largely be paid for by the feds. Once residency started, that coverage would end, so I'd have to moonlight to pay for it.

The only question I had was how to find the right psychoanalyst. In Boston there were a few psychiatrists who had a reputation within the professional community as being adept matchmakers, able to pair a potential analysand, as the patient was called, with the ideal analyst.

One such reputed matchmaker was an analyst named Max Day. Having no idea what to expect or how I was supposed to approach "the great Max Day," as some people in Boston referred to him, I impulsively called Information one day, got his number, and rang him up. He actually answered the phone himself. Somewhat surprised, I sputtered ahead and nervously told him what I was looking for. He said he had time to see me, so I booked an appointment to meet him in his home office in Newton.

With that one spur-of-the-moment phone call, I set in motion a process that would last twenty years, and actually, to this day, has not formally ended. It would cost me upwards of $200,000 and well over a thousand hours, if you include travel time. For such an enormous commitment of money, time, and energy, I did no research, unquestioningly took what the grapevine gave me as the man to see to get a referral, and with enormous hope, excitement, and faith burst into the unique, unpredictable, and enigmatic world of psychoanalysis all but blind.

Max Day's house was on a lake, and I thought to myself that first afternoon that he must make a lot of money doing this work to afford such a fine house in Newton.

The next thing I knew I was sitting in a ladder-back chair at the corner of his desk. Max sat behind the desk with a yellow legal pad and a pen in front of him. He wore beige polyester trousers, a blue-and-pink-checked short-sleeved shirt that looked as if it were bought at Sears Roebuck, and a garish, mismatched ultrawide tie that was one step short of being part of a clown's costume. He wore pale yellow eyeglasses, and his face looked like a pie. He might as well have been named Max Bialystock.

He looked at me and asked, "Why are you here?"

He could have been the man behind the counter of a deli asking, "Whad'll ya have?" *I'll have a psychoanalysis on rye, extra mustard, hold the mayo,* I felt like saying. But instead, I dutifully replied, "As I said on the phone, I'm here because I want to go into psycho-analysis and I understand you are good at matching people like me with the best analyst."

Max heaved a sigh, all but saying "Oy vey." He shrugged his shoulders and said, "Why do people keep saying that about me? I've never claimed any such thing. How could I? It's impossible. How can anyone predict what will happen when two people meet? I can't. Nobody can. If that's why you're here, I can't help you."

"But I told you on the phone," I protested.

"I must not have understood you," Max interrupted. "I would have said you've got the wrong guy if I'd understood you."

"So you don't match people with the right analyst?" I responded, feeling lost.

"Wouldn't it be nice?" Max replied wistfully, almost rhetorically. "I would do it if I could, but I can't. No one can. I can make a referral for you if you'd like me to, but that's just because I know a lot of analysts. They're always looking for business."

I felt confused. The man I was entrusting with one of the most important choices I'd ever make, whom to go to for my momentous, life-changing psychoanalysis, was telling me the choice was all luck of the draw. "Dr. Day, I was really hoping you could guide me on this. Picking the right analyst seems pretty important, don't you agree?"

Max raised both his hands, as if looking to heaven for help. "Yes, yes, of course it's important, it's very important, it's just that there's no good way to do it. I've seen people interview ten analysts before settling on one. So how could I possibly tell you who to go to see based on one meeting with you? You could hate me forever if I got it wrong."

"You've seen people interview ten analysts?" I asked, amazed.

"Even more! Sometimes I've seen them interview twelve, fifteen, like trying on suits at Milton's, and still not finding a single one they thought was right."

"That seems pretty ridiculous," I responded. *Of course it does,* I could hear the naysayers saying. *The whole process is bogus, like a cult.*

"Ridiculous is pretty common when it comes to the mind," Max replied. "You'll learn this, if you haven't already."

"Well, what should I do, then?" I asked. I felt like a big bubble was slowly getting burst. "I'd really feel more comfortable getting a referral from you rather than just blindly picking someone out of the Yellow Pages."

Max had a sweet, dimpled, old man's smile. "I don't think you'd find many of us in the Yellow Pages. But if we were there, that would probably be as good a way of finding the right analyst as any other. Still, you've driven all the way out here, you've come a long way, I assume you drove? You didn't take the T, I hope I told you not to take the T, since you've gone to all this trouble, if you want me to make you a referral, of course I'd be glad to. I just don't want you to get your hopes up too high."

Why the hell was he asking me how I got to his office, did I drive or take the T? Didn't he understand the urgency that I was feeling? "To tell you the truth, Dr. Day, my hopes were up very high, but the way you're talking, well, it's not what I expected."

"Life usually isn't, but I'm sure you've discovered that by now."

"Yes, I have," I responded, doing my best not to get angry or show frustration.

Max surprised me by saying, "I ought to get to know you a little bit if I am going to do this, don't you think?"

"But I thought you said it won't make any difference."

"It won't, but it's always good to get to know someone. It's a good thing to do. Not because it will determine who I refer you to—I'll probably just refer you to the first guy I can find who's in town and has time—but it's just a good thing to do."

"I don't know, Dr. Day, this is not what I'd expected. Are you testing me or something?"

"You've come a long way out here to see me, and I've disappointed you, but you're not getting mad at me. Do you have trouble with getting angry?"

"In fact, I do."

"Well, there's a start. We are learning already. Why don't you tell me your story?"

"You mean my life story?" I asked.

"Any story you want to tell me," Max replied. I actually do believe he'd have been just as happy if I told him the story of Goldilocks and the Three Bears. He was all about getting you to say whatever you wanted to say.

"I meant, wouldn't my life story help you in making the referral?"

"Probably not," Max said. "But telling me your life story will help you. Do you know how rarely a person gets to tell his life story to someone else? Most people die before they get around to doing it."

"Is that what happens in psychoanalysis?"

This time he played it straight. "Sometimes, but not always. I once analyzed a Holocaust survivor. The analysis took five years and not once, not one single time, did the man mention the Holocaust in his analysis. But it was a very successful analysis."

"What does that mean?" I asked, feeling excited again. "What's a successful analysis?"

"You tell me," Max answered. "What are you looking to buy?"

"Is what I am looking for for sale?"

"It depends what you're looking for," Max replied. "Most people are looking for happiness."

"I'd settle for that."

"That's not the usual result," he said. "Take pills if you want happiness."

"So what's the best I can hope for?"

"You tell me. What is the best you hope for?"

"I just want to be happy, and not in the superficial way you're referring to when you talk about pills. I want to get rid of the problems I carry around from my childhood."

"That would be a successful analysis," Max said, finally answering my question, before adding, "Just don't bet the ranch on getting it."

"Thanks for your honesty," I replied, instead of telling him what I wanted to say, that he was a real jerk for being so—what's the word?—realistic, pessimistic, dour, a downer?

"So now, please, tell me your story," Max said.

For the next hour—I don't know how much time he had blocked out for me, but I got the feeling I could have taken as long as I needed—I talked and Max listened.

He was a good listener. There's a reason people called him "the great Max Day." His quizzical, enigmatic style, while at times infuriating, and at times seeming manipulative (I guess it was intentionally manipulative), did cause me to think, and it did, as they say, "manage my expectations," which were, as usual for me, ridiculously, unrealistically, almost unmanageably high. Max proved to be an adept manager.

Also, as I told my tale, something about him made it not only easy to talk but also easy to go into dark places and not feel awkward, even though Max was a total stranger. He'd give off little cues—a slight turn of his head, a movement of his mouth, a slight shift in his gaze or a leaning back in his chair—that would reinforce my desire to go on, and invite me to go deeper.

I actually cried at one point during that hour, when I was talking about the bad luck my mother had and how she'd slipped into a life of drinking. He let me cry—which I almost never do—and said nothing, but slid a box of tissues toward me.

When I finished my story, he said, "Of course, that's just the beginning, isn't it?"

"There's a lot more. Yes, there is."

"There always is. That's why it's too bad more people don't take the time to tell their stories."

"Why don't they?"

"No one asks them to."

"Will you be giving me the name of an analyst?" I asked.

"Of course," Max said. "That's what you came here looking for, isn't it?"

"Yes."

"But that's not what you found, is it?" he said, writing something on his yellow pad, the first time he'd written anything our entire session.

His question put me on the spot, like on rounds when the attending is playing Guess What I'm Thinking. I didn't want to play a game, I was feeling pretty spent after telling my story, so I just said, "I don't know what you mean."

"You came here looking for a name, but you found something else," Max said.

"Well, I found you."

"It's not about me," Max said.

"OK, I give up. What did I find?"

Max put down his pen and looked up from his pad. He took off his glasses and wiped them with his ridiculous tie. "You found a door. A door you've been looking for for a long time."

A few days later I got a call from Max. "Dr. Hallowell, it's Max Day. I have a name for you."

I was standing in my living room looking out the window of the Brook House down at Brookline Avenue, Fenway Park in the distance.

"Do you have a pen?" he asked. "His name is Ed Khantzian." Max gave me Dr. Khantzian's phone number and told me he'd briefly spoken to him about me. "The main thing is he's in town and has time. I told you that would be the most important factor."

"Thanks, Dr. Day. I really appreciate it."

"Let's hope it works out well for you. Goodbye."

So my choice of analyst was governed by, of all things, availability. At least that's what Max Day told me. My fantasy—my wish—was that he thought long and hard about exactly what kind of person would be best for me. I wanted this decision to matter. I wanted Max Day to use all his brilliance and experience to plumb my depths and find the perfect person to conduct my psychoanalysis. I wanted to matter. You could say that in that wish my psychoanalysis began.

52.

When I met Dr. Khantzian, I did so with the grandest expectations imaginable. I hoped he could sort out my childhood and make me a happy person.

Our first meeting was, therefore, destined to be anticlimactic. I walked into his office on the third floor of a nondescript brick building next to Cambridge City Hospital. He was tallish, mostly bald, swarthy, wore glasses, and had a mustache. When I shook his hand I noticed he had a Boutonniere contracture, a hardened sheath of tissue around a tendon, on the pinky of his right hand.

I sat nervously in a chair opposite Dr. K. as he took my history, his legs crossed, writing notes on a yellow legal pad held in his lap. He didn't make many comments, just wrote down the facts, à la Sergeant Friday on *Dragnet*. With no drum roll or fanfare, we were off to the races. My eagerly anticipated psychoanalysis had begun.

After a few introductory sessions, he asked me if I was ready to "go to the couch." It sounded like such an intimidating proposal, like "go to the mat," or even "go to war." Could I maybe just go to the bathroom instead?

Of course, before sitting down I had immediately noticed that along one wall of his office he had an analytic couch, as they're called. They come in many shapes and sizes. His was about two feet narrower and a foot shorter than a twin bed. Some such couches, the ritzier ones, are covered in leather, but his was fabric, with a cushion at the head and a little rug at the foot so the analysand's shoes wouldn't soil the fabric. Before I arrived for my session he always put a clean paper napkin on the pillow. It reminded me of the dentist draping a napkin around my neck. Would I need Novocain for this as well?

When I was lying on the couch, he sat behind me in a reclining chair and put up his feet. Dr. Khantzian saw to it that he was comfortable, and he did his best to make sure I was comfortable as well.

Lying down on the analytic couch the first few times was discomfiting, but not nearly as awkward as I had expected it to be. I had expected to feel ridiculous, as if I were in a cartoon. But once I got over the unfamiliar sensation of talking to someone sitting behind me while lying flat on my back, I got used to it. Soon I felt completely at ease.

I took to it, a duck to water. I understood why Freud created this arrangement, as it indeed allows the patient to float along mentally, freely associating without the inhibiting force of eye contact and the other requisites of polite interchange between two people.

I could gas on and on without having to take into account the expression on Dr. Khantzian's face, his body language, his changing position in his recliner, his blowing his nose, or picking it for that matter. He could have been trimming his nails or doing the *Times* crossword for all I knew. Now and then I'd hear paper rattle as he turned a page in his spiral notebook. Each patient had a dedicated notebook—rather than a yellow pad.

He was unbelievably attentive. He never lost track of what I was saying, which amazes me to this day. I don't know how he did it; I couldn't have. Just his sustaining attention for that long was worth the price of admission.

I started my analysis in 1978, still an era of fairly classical technique, at least in the hands of some. My chief resident, John Ratey's analyst, one of the most respected and sought-after analysts in the Boston area, was of the classical variety. According to John, he hardly

260 Edward M. Hallowell

ever said a word. That was the old way. In fact, Freud actually stated that the analyst shouldn't listen too closely to the words the patient spoke, but rather sit with "evenly hovering attention" over the procedure.

The analysts who trained in New York City under the likes of Charles Brenner and Jacob Arlow were schooled in this blank-screen methodology. The idea was to keep "the field" clear of the analyst's self, allowing only the patient to introduce new material. Now and then the analyst could speak, usually offering an interpretation, but to speak often would be interfering with the process and deny the patient the chance to develop a relationship and live with it, even suffer with it if need be, before it got interpreted.

Dr. Khantzian had none of that, thank God. He interacted with me the way a normal person would, with three basic differences: First, I was lying on a couch and he was sitting behind me; second, the perpetual subject of discussion was me and my life; and third, I paid him. If I asked him a question about himself, he would give me a chance to wonder—thus developing my version of who he was, which is called transference. But if I really wanted to know, I'm pretty sure he'd have told me. I actually never did *really* want to know. I preferred to create my own version of him in my mind, to let the transference develop. After all, that's what I was paying him for, at least in part. It's what made the relationship unique and useful.

I never asked him personal questions, but over the course of my analysis I did glean, somehow, that he was happily married, that he had four children, that his father died when he was young, that he'd gone to medical school at Albany Medical College and done his residency at MMHC. During one of our sessions, he described himself as a "bald-headed Armenian from a shoe town." The shoe town in question was Haverhill, Massachusetts, some thirty-five miles north of Cambridge. I also learned he was interested in the treatment of addictions and that he and George Vaillant, another prominent analyst with similar interests, had heated disagreements on the subject. Vaillant had gone to Exeter and Harvard, came from an old New England family, and, on the face of it, might have seemed a better match for me as an analyst. But Max Day had chosen Khantzian. Max chose right.

Would a more classical analyst have been better for me, in that I would have had to bear more frustration, which would have led to growth and insight? I'll never know. What I do know is that my years with Dr. K. helped me enormously.

I went to see Dr. Khantzian four times a week. Each session lasted fifty minutes. The cost started at something like $80 per session in 1978 and went up gradually as the years went by. I can't recall how much it cost at the end, maybe $225 or something like that, but by then I was seeing him only once a week. I saw him intensively, four times a week, for about seven years, then tapered off, but never actually terminated.

"Terminate" is the verb analysts use for the process of ending the treatment. It is supposed to be a crucial phase of a psychoanalysis, during which the feelings come up that come up only when you know you will never see a person again. And, in a classical analysis, that's the rule. Once you terminate, you never meet again. Any hopes you might have of meeting again you address during the process of termination. We never bought into that model. We talked about it, why both of us didn't like that approach, and left it at that.

As my analysis went on and on, much longer than most analyses, some of my friends would ask me if I were cured yet. I would reply, "Yes, in the sense that a ham is cured."

To his face, I always called him Dr. Khantzian and he called me Ned. As time passed, behind his back, talking to friends or girlfriends about him, I'd refer to him as "my analyst," "Khantzian," or "Eddie the K."

His most outstanding quality was sanity. Of course, this is my assessment, and, since I had no contact with him outside our sessions, my assessment falls under the category of transference. But to my mind, he was a very sane and reliable man.

He was there for me, almost always on time, at every session, be it early in the morning or late in the afternoon, be it a sunny summer day or a nor'easter whipping down on us with snow and wind. During the sessions, I never kept track of the time and just kept talking until he said, "Well, it's about that time . . ." If I were to interpret that now, I'd say I was letting go of control and letting him take care of me.

Once it was "about that time," I'd do a quick sit-up from my supine position, put my feet on the floor, and stand, all while Eddie the K was getting up from his recliner and opening the door. I didn't shake his hand on my way out, not out of rudeness or irritation but out of a sense of not wanting to be chummy with him. This was not a friendship. I'm not sure what it was. But whatever it was, it changed my life in a way I do not believe anything else could have.

I honestly can't remember a single interpretation he made during my entire psychoanalysis, even though I am sure he made many. What I do remember is our relationship, and the trust I developed in him. He disappointed me a few times, in ways I won't take the time to go into because they were human errors that didn't really matter. All I'm saying is that he wasn't perfect, and I knew it. So did he.

Of course, he knew my *myriad* imperfections. He listened to me put myself down regularly. Poor self-esteem was one of my biggest issues. I was deeply insecure. I'd joke that if someone complimented my tie, I'd reply, "So you don't like my shirt?" I gave myself almost no credit for my accomplishments but instead dwelled on what I perceived to be my many shortcomings and failures. Not getting into any medical school the first time weighed heavily on me.

"But you know it's because you didn't take the sciences at Exeter, so you didn't have the foundation to put up top scores in college and on the MCAT," Dr. Khantzian would say. "And you have to give yourself credit for digging deep, getting a research job, and getting in the second time around."

"But I didn't get into Harvard Medical School," I replied. "That's where the top people go."

Dr. Khantzian laughed.

"I know, you went to Albany, so what are you going to say?"

"You got me," he said. "Don't you think you're being a bit of a snob?"

"That's what going to Exeter and Harvard does to a person, at least to me. Once I didn't get in anywhere, it's like I fell off the train to the top and I could never get back on. I'm just second rate. I better get used to it."

"I think you know that doesn't make any sense."

"But it's how I feel."

"OK, let's stick with that and see where it takes us."

That's the huge advantage of psychoanalysis. You have enough time—boy, do you ever—to stick with something and see where it takes *us*, us being a key concept, because, at least the way Dr. Khantzian practiced psychoanalysis, I was not alone.

Where it took us was not to stunning insights, but rather to an ongoing sharing of the problem. Because of the problems that surrounded me when I was growing up, for which, I should add, I blame no one, I paid the price of being insecure and low on self-esteem as an adult. To fix those problems, I needed someone to stay with me for a long time in a way my parents hadn't been able to.

Dr. K. couldn't repair my problems just with clever interpretations or insights, but he could help me, if not "cure" me, by hanging in there and keeping the relationship going until I felt good enough to leave.

As my relationship with Dr. K. evolved, and my psychoanalysis progressed, I found myself regretting less and less and finally not at all my stupid stuff like not getting into Harvard Medical School, and letting myself feel sad about my important stuff, like my dad's mental illness and my turbulent years in Charleston. Instead of pushing the pain away, I let myself feel it, as Dr. K. listened and made comments, which, in a process I don't understand, actually dissipated the pain over time. I came to feel a lightness I'd not felt before, at least not since I was a very little boy.

Max Day was right to downplay to me the results of a successful analysis. After he heard my story, I imagine he knew it was going to be a long haul for me no matter what analyst he referred me to, which I believe is why he kept telling me availability would be the deciding factor in his referral. He didn't want me to believe anyone had the magic touch. As it turned out, though, Khantzian sort of did.

Of course, we talked about my mother. She was drinking heavily, calling me when she was drunk, and I would listen for a while before saying goodbye. Even before I started seeing Dr. K., I had cut myself off from the deepest ties I felt for my mother. I was no longer the little boy who'd prayed every night at Fessenden that she could be happy, nor was I the distraught college student who worried constantly over her drinking and sadnesses.

The deepest part of me must have known—but never articulated—that it was basically a choice between her and me. I could live the rest of my life for her—a losing battle, without any doubt—or I could cut the ties and try to fashion a life of my own.

I didn't cut off contact or anything like that. I was there for her right up until she died. But I did stop worrying so much about her. I did not consciously decide to do that; it just happened. I still cared for my mother, I still loved her, but I stopped letting her happiness determine mine.

I still have not officially terminated with Dr. Khantzian. One of us will die before that happens. A classical analyst might object to this, but if they knew the whole story, I don't think they would.

#

I t is love that brings them to us and love that sends them away," said Elvin Semrad, who, soon after seeing a patient and then frying an egg in a skillet on the hot plate he kept next to his desk in his office, fell dead of a heart attack. Although that was a year before I got to MMHC, Semrad influenced me more than any figure in the field of psychiatry through the oral tradition he left behind.

In its heyday MMHC, both a state hospital and a Harvard teaching hospital, stood as the premier training program in the country, and Semrad, by training so many psychiatrists who would go on to other parts of the country to lead countless other training programs, shaped the thinking and practice of thousands of psychiatrists.

"A hayseed from Nebraska," as he called himself, Semrad was a wildflower for sure in the fiercely intellectual, competitively cerebral Boston psychoanalytic community, which was deeply rooted in the Viennese tradition of Freud. He no more fit in with them than a square dancer in a ballet. But Semrad had an uncanny and undeniable genius for connecting with and understanding people, a talent that many tried to emulate but few could replicate. Even the brainiest analyst had to admit Semrad had a gift few, if any, of them could match. He influenced deeply a generation of psychiatrists, including, indirectly, others like me.

Today, I still feel him with me, this teacher I never met, a guiding star in a field that always has been, and likely always will be, trying to find its bearings, until, as Freud himself envisioned, it deepens its roots sufficiently in basic science to get swallowed up as a subspecialty of medicine, neurology, and neurosurgery and disappears forever.

Today we are closer to Freud's dream than ever before, which is mostly to the good. We have brain scans of various types that have enhanced both our understanding of how the brain works and our ability to diagnose. We have an array of medications that can relieve suffering as never before, and we have standards of diagnosis that are becoming ever more precise. And we've added to our therapeutic arsenal many new interventions, like surgery for deep brain stimulation; vagal nerve stimulation or transcranial magnetic stimulation to alleviate intractable depression or obsessive compulsive disorder; old drugs with new applications, like ketamine for depression or 3,4-methylenedioxymethamphetamine (MDMA, called on the street Molly or Ecstasy) for post-traumatic stress disorder (PTSD); eye movement desensitization and reprocessing (EMDR) also for PTSD; and dialectical behavioral therapy (DBT) for a host of conditions, especially borderline personality disorder, which for decades was so intractable.

When I started my residency in 1979, psychiatry, once the field of "talk therapy," was focusing more and more on neuroscience. Coming of age in a field that was finally freeing itself from the rigid orthodoxies of traditional psychoanalysis and beginning to embrace the immense benefits of biological interventions, I was lucky enough to learn from both worlds, two groups of professors, one slowly fading but wise nonetheless, the other coming on strong, powerful tools in hand, and about to create a new orthodoxy. The brain was supplanting the mind before my very eyes.

As always, the fanatics would do damage. We need an all-inclusive psychiatry, not a field divided by turf battles amongst egoists. Turf battles make me think of one of my favorite prayers: *Lord, help me always to search for the truth, but spare me the company of those who have found it.* At MMHC, we had teachers (called supervisors) who were analysts or had grown up in the analytic tradition, but the staff had no fanatic Freudians. Quite the opposite. There were people like Les Havens, Tom Gutheil, Irv Taube, and Doris Benaron who, while psychoanalysts by training, were humanists by disposition. They favored the way of Semrad: Go where the patient is. Follow the feelings. Use empathy to guide you.

And the leaders of brain science at MMHC, the advocates of the biological interventions—Allan Hobson, Joe Schildkraut, Richard Shader, Carl Salzman—also insisted we learn and practice the skills of forming an alliance, understanding the patient's life, and staying with feelings even as we were making a diagnosis and prescribing a medication or other biological intervention.

For us residents, what could have been turf wars instead became the sharing of knowledge and ideas, each camp offering what it had and eager to learn from the other. That may sound too good to be true, but it's what we, or at least I, experienced every day in my training. It couldn't have been better.

But for all the theories and ideologies, there was—and still is—only one group of *patients.* There were and always will be special cases whom we doctors came to know and love, in a manner unique in the varied world of loving relationships.

These patients didn't give a hoot whether psychoanalysis, biological psychiatry, interpersonal psychiatry, or colonic detox was the current craze. They simply presented themselves to us, in full puzzlement, on the off chance that we could somehow make a deal. They gave us their stories, their selves, and in so doing taught us in the best way possible. In return, we applied what knowledge and methods we had to try to improve their lives. For trainees, that was a great deal. For the patients, well, they drew the short straw, but they got the best we had to give.

54.

The morning we arrived at MMHC, all of us new first-year residents gathered in the modest hospital library, replete with oak tables, monogrammed Harvard chairs, thumbed-over classics, and obsolescent tomes. Greeting us were Dr. Miles Shore, Bullard Professor of Psychiatry at Harvard Medical School and superintendent of MMHC, along with our training director, Dr. Leon Shapiro, a tall, savvy psychoanalyst (actually, a training analyst, which is like a knight or a Jedi in that guild). Also present was Dr. Leston Havens, one of the most inspiring and learned mentors I would ever have.

Havens, a man who did his best to empower residents not to feel intimidated by the brass, showed up seemingly fresh from a walk along the Charles River in Bermuda shorts and sandals, while Shapiro and Shore wore coats and ties, as did most of the other faculty there that day. They are a blur now, but then I looked at them as the guardians of the Holy Grail.

Once each senior faculty expressed a few words of welcome that morning, it was announced that the first night on call would go to one Dr. Hallowell.

Heading up to the fourth floor, I took the shaky, broken-down gray metal elevator. Lots of things at MMHC were gray metal, shaky, and broken-down, including desks, beds, chairs, and tables. I'd spend the next year in this ward—innocuously called Service One, one flight above Service Two—discovering that the cultures of these two services were as different as Dionysus and Apollo. Fittingly, I was on the Dionysian service.

Instead of a stethoscope, I now carried a set of keys in my pocket. I would later write an essay "On the Transition from Stethoscope to Keys," which Dr. Shapiro ridiculed as stupid and simple-minded, and he advised me not to enter it in the contest I had written it for.

As his words took me back to sparring with Uncle Unger, of course I joined the joust. "Maybe I don't want to write boring, psychoanalytic articles no one will ever read," I replied.

Shapiro smiled, as if he liked my spunk, rather than slapping down this rookie, first-year resident, for which I will always be grateful. But I did take his advice and did not enter the paper in the contest. However, a year later I submitted another paper in that contest, on the use of poetry in doing psychotherapy with a schizophrenic patient, and it won first prize, the Solomon Award, beating out the entry from Dr. Shore, who wrote on an innovative method of caring for the chronically mentally ill. That day I enjoyed what psychoanalysts—both Shore and Shapiro were psychoanalysts, because in their day becoming one was de rigueur—call an Oedipal victory. I had slain the father. This is often followed by anxiety and guilt, or by confidence and swagger. I felt all four.

Now, instead of scrubs and a white jacket, I wore what most of us wore, casual street clothes. We didn't look like doctors. A few of us wore jacket and tie, which Dr. Shore applauded, but most took advantage of the freedom to dress down, which seemed more in keeping with the physical condition of this state hospital. It was a dump, but a dump patients loved and I came to love as well. It was a glorious dump, if you ask me. A true rag-and-bone shop of the heart.

Upon arriving on the ward, the five of us new residents and the one psychology intern made our way to our offices, which John Ratey, our chief resident, had randomly assigned. The patients were milling around with obvious curiosity, like fish circling, checking us out, as were the nurses and attendants on duty. They'd awaited our coming, not knowing what kind of crop they'd get. The day before, they'd bidden farewell to the previous year's residents, and now they were eager to see this new troupe of doctors in training. It made sense that they were eager, since in 1979 more than a few patients stayed at the facility for over a year. Some patients would be with us for a long time.

However, when proceeding to my new office, I didn't notice the patients right then. I had in mind another purpose altogether. I'd brought with me a poem I'd typed and framed. Dating back to college, I'd planned to hang this poem on the wall of my first psychiatry office, if I ever got one. Now that I was here, standing at the door of that office, I felt I should invoke some blessing. I paused and waited two beats. It's the easiest way to make any moment special. Just wait two beats. Stop time.

When I opened the door, it was love at first sight. No matter that the chairs were naugahyde, some with rips and tears, the desk was dented gray metal (of course), the windowpanes were filthy, there was a hole the size of a grapefruit in one wall with hairs of plaster hanging out of it, and another hole, this one the size of a bullet, in one of the windowpanes. I was just grateful there were windows, and four of them to boot, each with eight rows of four windowpanes. I was home at last.

My view looked out onto Fenwood Road. Up the hill I could see the Peter Bent Brigham Hospital where the men and women who had been my fellow interns were at that same moment showing up for their first day of medical residency. I looked out my office window and gave them all a wave. What different lives we would now embark upon.

Turning away from the window, I looked for a place to hang my poem. I'd brought a small hammer and hook with me, so all I needed to do now was to pick the spot.

I chose the wall directly opposite the door, so I would see the poem every time I walked in. It would mean nothing to anyone else, except possibly those few who might get close enough to read it, but for me it set the stage for the coming year and for my entire career. It was W. H. Auden's poem "Musée des Beaux Arts," in which Auden uses Breughel's painting of Icarus falling into the sea as a model for human suffering. In the poem he notes "how everything turns away quite leisurely from the disaster." Even though the sailors on the ship had seen "something amazing, a boy falling out of the sky," they took no notice as they sailed calmly along to wherever they were headed, the disaster, if even noticed, quite forgotten.

I pounced on that poem in college and turned it into an emblem for my future work. Not only did it speak to the lives of the people I would treat, it spoke directly to me, to the loneliness and sadness I felt growing up, which the world couldn't notice because it had somewhere to get to. Of course, my suffering was puny compared to that of the tens of millions of children who *really* had it bad, some of whom I would meet in the coming years, but my suffering was, in its own unimportant way, what *I* had to deal with.

The patients at MMHC had it much worse than I could have imagined. I had no idea how tough the lives of the patients I would meet over the coming years really were. These patients were poor *and* crazy. There is no group more overlooked, if not outright despised, than the indigent mentally ill. It's bad enough to have no money, but to be out of your mind as well? As far as the general public is concerned, you occupy the place lepers used to take.

MMHC served as its own kind of leper colony. Politicians wanted no part of it, cutting its budget whenever they could. The hospital's neighbors wished it weren't there, and even the Harvard medical students who rotated through usually held their nose until their rotation blessedly came to an end. Most preferred to do their psychiatry rotation at McLean, the posh hospital where the well-heeled and oftentimes famous went for treatment.

I chose the dump. As I put my feet up on my dented desk and tilted back (yes, I even had a chair that *tilted*), I felt I'd found my own special paradise.

Perhaps because I'd always felt like an outsider, I felt instantly at home at this stopping place for those who did not belong anywhere: the marginalized, the misfits, the estranged. Misfits, take heart. We have a home. Welcome, O life! I go to encounter for the millionth time the reality of experience and to forge in the smithy of my soul . . .

As all my compatriots left Service One to enjoy their day off that Sunday, I entwined my fingers behind my head, leaned back, stared out the 128 windows and through the one bullet hole, and thanked my lucky stars for landing me at MMHC.

In years to come, I would have thanked God, but back then I was on leave from God. I thought about God, as I had done ever since my days in Charleston, but I didn't pray regularly anymore or ever go to church, not since required church at Exeter and the sermons of Fred Buechner. Unless you couldn't shake your Catholic upbringing or you were an Orthodox something—Jew, Muslim, Greek, Buddhist— medical training typically instilled extreme doubt, if not a sturdy atheism supported by science and logic.

We saw too much senseless suffering to get past the obvious question everyone asks when they first learn how to question: How can there be a God if all this horror happens every day? Or, as my best friend in medical school, Steve Bishop, told me Martin Buber wrote, "If God is good, He is not God; and if God is God, He is not good."

I had no idea what my years at MMHC would lead me through, any more than my recent friends up the street at the Brigham knew what their years in medicine would lead them through, what changes, what joy, or, sometimes, what hardening of the heart.

I thought of them, and imagined what they would think if they saw me now at MMHC. "What are you doing here in this dump?" Mark would ask in disbelief. "What a waste of talent. It's not too late! Come join us!"

I was committing to a stranger road, a road not widely respected by doctors or the general public, a road with few clear road signs or dependable destinations.

In coming to MMHC I was joining *my* people, the wildflowers of this world, the people I grew up with, the people I understood in ways that most people can't, people with whom I found it fun to hang out, as I did for hours on end in the day room, people most of the world fears and dislikes, but people whom I liked. I knew I could connect with them. I hoped I could even help.

55.

My second day of residency, I sat beside Professor Paul Stein in the day room. He was a tall, gaunt man with a tobacco-stained four-inch beard. Head nurse Linda had told me his name but that was all I knew about him. He was a patient at MMHC but not my patient. He was just a person I wanted to get to know. I was out in the fields again, with the people this world didn't cultivate or care for.

Standard practice is to read a patient's chart before meeting them, to get background and context so as not to unknowingly stumble into sensitive areas. I didn't want to do it that way. I preferred meeting a person without preconceptions. Why shouldn't *I* meet *him* on the same terms he was to meet me? As in this case, many of my instincts went against regular procedures. The beauty of training at MMHC is that the powers that be encouraged innovation rather than always operating by the book, as long as what you did was safe and legal.

So I met Professor Stein much as he met me, with no prior knowledge. Because my father had been saved by a young doctor's taking a fresh look at my dad and radically changing his diagnosis and treatment, I knew on a personal level how wrong psychiatric diagnosis can be, how rigid treatment plans can become, and how valuable *not* knowing what's been concluded by others can prove to be. Meeting Professor Stein cold provided me my one chance to see him through my eyes only.

"Sit with the patient" had been Semrad's command, and John Ratey, my chief resident, had also advised us to at first just sit with the patient. So there we sat: Professor Stein and me, him staring straight ahead, both hands on the arms of his high-backed chair, and me, in a folding chair next to him, turned forty-five degrees toward him so I could either look at him or join him in staring straight ahead. I was biting my tongue, holding back my desire to pepper him with questions.

We sat in silence. He took out a cigarette and lit it. More silence . . .

I was trying to get used to sitting next to someone and remaining quiet. People don't generally do this in polite society. But this was not polite society. Here I was, freed of all the tests, technologies, and busyness of regular medicine, charged only with the task of connecting with another person. And so we sat.

Finally, as if I'd held my breath as long as I could, I had to break the silence. "You've been here a while," I said. Professor Stein just kept staring straight ahead.

I began to wish I'd read the chart. *Could* he speak? Was he mute? What meds was he on? What diagnosis did he carry? Did he have family? Was he actually a professor, or was that a nickname?

On and on we sat. He leaned forward and tipped his ash into the red plastic ashtray on the "coffee table" in front of us. I put it in quotes because so much at MMHC was ersatz. The patient "library" was a sitting room with shelves upon which happened to be a random assortment of tattered paperbacks and a few broken hardbacks. The "solarium" was a room that once had a skylight but now was sealed over. The "phlebotomists" were a couple of guys who wore T-shirts and carried toolboxes full of the equipment needed to draw blood. The "TV room" was a smallish room that had a semifunctional tabletop TV with rabbit ears. The "linen closet" was a closet stuffed with towels that looked to be older than I was. "Housekeeping" was a group of state employees who were spooked by the patients and spent as little time on the ward as they could. And the largest room, the "day room," sported a motley assortment of "easy chairs" with steel frames and naugahyde cushions, rickety folding chairs, one "sofa" that had its stuffing popping out all over, and at one end of the room a "fireplace," which, of course, was bricked over. Sometimes even I felt ersatz, having just days before been practicing real medicine as a real doctor in a real hospital, only now to find myself entering a field that the rest of the world joked about.

I was doing my best imitation of a real psychiatrist, sitting next to a man who was, without doubt, real. Whoever he was, whatever his story, whatever was going through his mind as Professor Stein smoked his cigarettes and sat next to me, it was real.

I began to feel annoyed with John Ratey for leaving us with such scant instruction. On the medical wards, they told us exactly how to insert an IV, how to do an EKG, how to perform a lumbar puncture, how to insert a catheter or an NG tube. And if ever we felt lost, they told us to always ask for help; seek counsel from a resident, a nurse, or an attending. Help was *always* nearby and we should *never* proceed without knowing what we were doing.

Ha! Here I was with no idea what I was doing, other than my chief resident's advice to "sit with the patient." I'd been told to "figure it out." There was no rulebook to follow, and as much as that's what drew me into psychiatry, now I was feeling frustrated.

I don't know how much time elapsed before the professor broke the ice. It felt like an hour—but he'd had three cigarettes, so maybe it'd been twenty minutes—when Professor Stein finally asked, "Do you play bridge?"

56.

"Why is he called Professor?" I asked Linda, the head nurse. We were both sitting at the nurses' station, one of the rare moments you could find Linda sitting down. She was drinking a coffee she'd just poured herself from the stale pot on the table next to the chart rack.

"Because he *was* a professor, of physics, at MIT, supposedly one of the youngest to get tenure. But that was a long time ago. He had his first breakdown when he was in his twenties, I think. I should know 'cuz I've heard his case presented so many times. He's had more residents take care of him than just about anyone here, maybe except Freddy. He's been in and out for decades."

"Does he have any family?"

"No one that he's still in touch with, at least as far as I know. Way back, some MIT faculty would visit him, but they stopped coming a long time ago. In case you haven't noticed, friends and families give up on these people. It's sad, because the patients who do best are the ones friends and family stick with. If you want my advice, when possible, involve the family. It makes all the difference."

"He's not my patient, I was just trying to get to know him."

"Oh. Well, good for you. Another piece of advice, if you want it: Spend more time with the patients than anything else. Lots of residents spend all their time hanging out here in the nurses' station or in supervision or in seminars. That's fine, but your real learning comes from spending time with these people." Linda motioned to the patients in the day room, each sitting in their own particular world. "Lots of residents say they don't have time to do that. My advice is to make time. They really are your best teachers."

Linda was a petite Italian beauty who understood patients better than just about anyone else at MMHC. I always wanted her advice. Everyone loved Linda. It would be so easy to fall *in* love with her.

I got back down to business. "I have Kenny Luongo on my list of patients assigned to me. Has he been here a while?"

"Not too long," Linda said. "He was refusing meds when the old residents left, so you'll be getting some pressure to get him on meds."

"Why won't he take them?" I asked.

"Well, handsome, for one reason, he's crazy." Linda gave me a little punch to my arm.

"Can he talk about why he doesn't want to take meds?"

"Sure, but it won't make sense. Give it a try, though."

This time I did look at the chart. Flying blind had a theoretical appeal but practical shortcomings. Kenny was twenty-four, admitted two weeks before I arrived at MMHC, and had been under the care of a resident I knew and respected. Kenny was refusing medication because he thought it was dangerous. The last progress note said Kenny was "incoherent and psychotic" and unable to have a meaningful discussion about his care. The advice was to appeal for a court order to medicate.

I had to ask Greg, one of the attendants, to point out Kenny to me in the day room, as I'd never met him.

"He's over by the window," Greg said. "He can stand and stare out that window for hours. Don't come up from behind him. He's like a horse. He scares real easy and doesn't like it one bit."

I sat in a chair about six feet to Kenny's left, well within his sightline, should he look my way, but to the side enough that he didn't need to look at me if he didn't want to. A basic lesson we were being taught about paranoid people was never to confront them. Leston Havens called this the counterprojective technique. Since projection is the chief defense mechanism in paranoia, you counter that process by siding with it, not getting in its way.

The natural response to a paranoid statement is to argue with it. If a friend says, "My boss hates me," you tend to say, "How do you know?" or even, "Are you sure you're not being paranoid?" But if a truly paranoid person says that his boss hates him, it's better to reply, "That must make working for him really tough," or something empathic like that. The goal is to create an alliance, at least in the beginning, not to ascertain the truth.

I sat a while to see if Kenny might look my way. After five minutes or so, he did. Then he looked back out the window. I waited longer, not looking at him but staying put. After another five minutes he pulled a cigarette out of his rolled-up sleeve, then bounded over. Standing above me, shifting from one foot to the other, he hurriedly asked, "Hey, man, you got a light?"

Damn. I didn't. I would learn to carry matches. "No, but I can get you one."

"Fuck off," he said, and returned to his spot. I went and got a book of matches just the same. I went back to my chair, saying "I have matches if you still want them."

"Fuck you," he said, holding the unlit cigarette. Then he walked back over, cigarette in his mouth, and leaned down for me to light it.

I struck the match. Kenny accepted the light, and took a long drag. "Thanks, man."

"Kenny, I'm Dr. Hallowell. I'm your new doctor."

"That's cool."

"Wanna sit down?"

"No." He stayed standing, smoking. I remained sitting.

"Nice to meet you, then. Maybe we talk again another time?"

"No chance," Kenny said. Note to Ned: Paranoid people do not like closeness. This was a hard lesson for me to learn because my natural tendency was the opposite. Having grown up in my enmeshed family, I left doors unlocked, lent out everything I owned, didn't balance my checkbook, and reflexively trusted *everyone*. I had to learn that many other people trusted no one.

A few days later, I approached Kenny again. He was at his usual perch. This time I had a light to offer him, which he accepted. This time he sat down in a chair next to me in the day room.

We sat side by side, looking out at the room and not speaking for a while. Then I asked if he was being taken good care of. He nodded. I asked if he was eating all right. He nodded. Since the nurses' notes continued to report that he was psychotic, I felt the need to bring up the issue of medication, but I didn't want him to freak out.

"By the way, Kenny, there's some medication that might be helpful to you. Could we talk about that?"

"No."

"Well, of course you don't have to talk about it if you don't want to, but could you maybe tell me *why* you don't want to?"

No response. I began to think Kenny wasn't going to reply at all when he blurted out, "It's personal." Then he waited several seconds before adding, "It's so personal even I don't know why."

Kenny may have been psychotic, but that was one of the most astute remarks I'd ever heard.

At that time, the issue of forced medication was in flux. The era of doing psychotherapy with psychotic patients, championed at MMHC by elder wise men and women like Semrad, Max Day, and Doris Benaron, was slowly coming to an end. I loved talking with psychotic patients, but I also knew that talk could not fix their psychosis nearly as well as medication could. Still, a human relationship would help them, no matter what.

This was liberal Massachusetts, this was Harvard. This was the epicenter of patients' rights, where patients were "allowed to rot with their rights on," as Tom Gutheil put it. It was odd and, to my way of thinking, unfortunate that so many people championed the right of a patient to remain psychotic, as if that were a precious right to be preserved at any cost.

Almost none of us—liberal, conservative, or disinterested—who were charged with taking care of these patients saw it that way. It had nothing to do with politics, it had only to do with what we believed was the best possible care we could provide. Letting a person stay in a psychotic state was, in our view, like allowing a patient to writhe in pain without intervention.

Psychosis hurts. As one of my patients said to me, "I take such a shellacking from the voices." Hearing voices inside your head—intrusive, unwanted voices called auditory hallucinations—is extremely upsetting. One of the many reasons that the auditory hallucination I had at age eleven was unique is that it didn't upset me in the least. Hearing voices is torture for most people.

And when the voices start telling you terrible things—like that you are being controlled by a machine at Logan Airport, or that the CIA is following your every move, or that your food is poisoned, or, worst of all, that you should kill a specific person or yourself—then you want to hold your head and scream, which many psychotic people do.

Those last are called "command hallucinations," voices that tell you to do certain things, the most awful being to kill other people or yourself. To make matters worse, for reasons I don't understand, it is extremely difficult not to obey those voices. The person hearing the voices is not able to say what I was able to say the one time I heard a voice inside my head, "Gee, that's interesting, let me give it some thought."

Instead, the person feels compelled to obey. They hate and fear the voices to the point that sometimes they commit suicide to escape the voices. So the development of medications that could take away the voices was a major step forward in treatment.

But what was I to do with Kenny? He had his personal reasons—unknown even to him, as he so poignantly expressed—for refusing antipsychotic medication. I could try to use logic with him, or I could go before a judge and obtain an order to medicate Kenny against his will.

Going before a judge wasn't as cumbersome as it sounds, because judges came to the hospital on a regular basis to hear cases like Kenny's. But after you obtained the order from the judge, what happened next wasn't always pretty. If the patient wouldn't agree to swallow the medicine, he'd have to be held down by attendants and have the medication injected. Even though it was legal and in the patient's best interests, it felt like a violation, which is in part why the people who stood up for patients' rights objected so vehemently.

When it was an emergency, we didn't have to go before a judge. But in Kenny's case, a judge would have to hear the case. It was extremely rare—in fact, I can never recall its happening—that a judge did not go along with what we recommended.

In the back of my mind, I never forgot my father telling me how much he feared and hated the insulin shock treatments he was subjected to. He said they were far worse than electric shock treatment.

"I didn't like the electric shock," he said, "but the insulin shock, that was a bastard." We were driving in a car when he told me this, I don't remember when or where. It was one of the few times Dad ever talked about his treatments. "They'd tie me down, then inject me with insulin, and I'd spiral off. It was like I went straight to hell. I became incredibly strong, I fought like mad, I wanted to rip people's heads off, it was pure torture. They don't do it anymore, thank God. Whoever invented it was a mean sonofabitch, I can tell you that."

Often, when I saw a patient being held down and injected, I thought of Dad. Then I'd tell myself that this was totally different, this patient was out of control and needed help before he hurt himself or someone else.

We did end up medicating Kenny against his wishes. Faced with the prospect of swallowing a pill or being held down and injected, he chose to swallow the pill, cursing us, especially me, all the while.

But unlike my dad, I do not believe Kenny will ever regret the treatment he received. The medication cleared his psychosis in days, so that he became able to converse meaningfully with me, with the staff, with other patients, participate in groups, exercise, and make progress.

By the time I finished my training I'd seen enough examples like Kenny to feel almost no compunction about seeking forced medication treatment. Indeed, I would have felt guilty if I hadn't sought it. The potential benefits are huge, and if it is done properly, the risks are very low. The risks of not treating far, far exceed the risks of treating.

57.

Division by zero," Rodney said to me while straightening his vest. We were sitting in the day room in two chairs next to each other. It was the middle of the afternoon, and surprisingly quiet, with most of the patients in groups or out on walks. Rodney was not my patient, but I enjoyed chatting with him. "It makes no sense that they call it undefined," he said. "If you can multiply by zero and if division is the inverse of multiplication, why in the world can't you divide by zero?" Rodney always dressed in as dapper a fashion as he could, retrieving clothes from the hospital's lost-and-found and collection of donated clothing. He described himself as "the best dressed lunatic in Boston." He meant it affectionately, as he held lunacy in high regard. "Most geniuses are crazy, you know that don't you?" he said to me once.

"Rodney, have you sent off your paper?" I asked. We had an ongoing conversation about his paper. I wanted to encourage him, but at the same time help him be cautious.

"Yes, of course. It will get rejected again, but I will keep trying until I prevail."

"Maybe try a less illustrious journal than *Science*? Is there some journal that likes offbeat articles?"

"My good doctor, the point I am trying to prove is an illustrious point! It's not offbeat. I want to give division by zero the place it deserves not only in mathematics but in the human experience."

"What do you mean, Rodney? I don't think many people think about division by zero as they go about their daily lives. It's not really a part of most people's human experience."

"Of course they don't. Any more than they think about splitting the atom, but that doesn't mean it doesn't have a profound impact on how they live."

"But Rodney, how does division by zero affect anyone's life?"

"Because it is undefined! What a ridiculous category for something as fundamental as that! Remember, there was a time when the smartest, wisest people in the world *knew* the world was flat! Where are they now? What do we 'know' now that is as fundamentally wrong as that? At least we call division by zero undefined, instead of calling it flat."

"Well, I just don't want you to get your hopes up, then have them dashed."

"Oh, Dr. Hallowell, you are such a kind, young doctor. But you have much to learn about life. Life will teach you that it's all about getting your hopes up, then having them dashed, then getting them up again."

I thought of Samuel Johnson's line, "Life is a progress not from pleasure to pleasure but from hope to hope."

"I guess you are a wise man, Rodney," I said.

He straightened his vest again. "Why thank you, my good sir."

On any given day, there were about forty patients on Service One, each quite an individual. Diagnoses like schizophrenia and bipolar disorder were common, but they told you little about the actual person. One of the great advantages of training at MMHC at this time was that we were all encouraged to get to know each patient as a person, not a diagnosis. This came from Semrad, who used to say, "The patient is the textbook. Spend time sitting in the day room with patients, instead of the library."

Eric Kandel, who trained at Mass Mental and went on to win a Nobel Prize in Physiology or Medicine, took strong issue with that statement, scorning it as anti-intellectual. But I think Kandel and

Semrad represent opposite sides of the same coin: the coin being the road to truth, in this instance the truth of the patient. You might say Semrad took the Shakespearean road, Kandel the Newtonian. In any case, both were great teachers. (Kandel still is.)

I never thought of Rodney as schizophrenic, which was his diagnosis, or as a lunatic, which was how he diagnosed himself. He was simply Rodney. To paraphrase a well-known line, there's far more in human nature than was ever dreamt of in the diagnostic manual.

We needed diagnoses to guide treatment, to do research, to communicate with each other and families and the world, of course, but I still always thought of Rodney as "the best dressed lunatic in Boston" rather than the schizophrenic in room 4.

#

Sitting at a table in the day room, playing bridge with Professor Stein and two new patients, I felt the presence of someone behind me.

"Melanie, are you peeking so you can give me some tips?" asked Professor Stein.

"No," Melanie said.

When I looked over my shoulder, Melanie backed up a few steps. She was a younger woman but with the worn look of a person who's been taking antipsychotic medications and battling mental monsters for many years. I didn't know her well, as she was not my patient, but she had an innocence about her that I'd come to learn many people with chronic schizophrenia shared. The ordinary citizen often fears chronic schizophrenics, but the average banker is far more dangerous.

"Dr. Hallowell, I would like to borrow two dollars from you," Melanie blurted out, then backed up another step.

I reached into my hip pocket for my wallet and looked inside. It was empty of cash. "Gee, Melanie, I don't have any cash. Sorry." The other three bridge players were listening closely to this conversation, as it was not customary for patients to ask doctors for loans, nor for doctors to check in their wallets in response. If I'd thought about it, I probably would have realized that not lending patients money was another unwritten rule I should have known about.

Instead, I blundered on ahead, guided by instinct and an inbred tendency to be polite. "Do you have privileges to leave the hospital?"

"Yes," Melanie said.

"If I were to give you my ATM card, could you go across the street and withdraw some money from my account? You'd save me a trip, and you could keep two dollars for yourself."

Melanie broke into a big, sweet smile, revealing her dire need of dental work. "You'd do that?"

"Yes, I would do that. I believe I can trust you."

"Oh, you can. I will be sure to be very careful. You can trust me for sure."

I took a scrap of paper from my wallet—I always have scraps, as scraps are one of my main tools of organization—and wrote down my password for Melanie, which I then handed to her along with my bank card. "Take out fifty dollars, OK?"

She received the card and the scrap of paper with two hands, then carefully put them into her jeans pocket. "Thank you, Dr. Hallowell," she said, looking flustered, then made her way to the door and off the ward.

"Doc, are you crazy?" Professor Stein asked. "She could empty your account and get on a bus to Kansas. You may never see her again."

"I don't think she will do that, Paul. I think I can trust her."

"That's nice of you, but do you *know* that you can trust her?" one of the other players asked. "She's here for a reason. She's crazy."

"Good point," I said. "On the other hand, I think she will rise to the occasion."

"But you don't know that," the player replied. "And what will you do if all of us start asking you for money? Have you thought of that?"

"I actually hadn't thought of that. I hope you don't."

"We shall see," Professor Stein pronounced. "Now, let's get back to the game!"

As I considered my next bid, I realized I had just made what pretty much everyone would call a huge mistake. I anticipated telling my supervisors about it and being read the riot act. "Are you nuts?" I imagined them saying. "Have you never heard of boundaries?"

Jumping to three hearts in response to the professor's opening bid of one heart, I quietly asked myself why on earth I had given Melanie my credit card. It was impulsive, not at all the result of deliberation. Why would I cross the boundary between staff and patient so blatantly?

An answer came to me, which I hoped was not merely a defensive rationalization. I thought the patients in general were capable of handling far more responsibility than we gave them. Melanie, in particular, who'd earned privileges to leave the hospital, would benefit from being given my trust; it would amount to a therapeutic intervention, not a foolish crossing of boundaries.

And why do we have to be so insistent on what separates us from our patients? I mused to myself, watching the professor take us to six hearts without even asking for aces. We're forever putting ourselves in charge of them in the name of protecting them and encouraging them to earn the right to leave the hospital and set their own boundaries. But, Ned, remember that in your upbringing the only boundaries you ever learned about were the ones studied at school in Geography. In Psychiatry, they matter a whole lot more than you realize.

As I carried on this internal discussion and watched the professor make our small slam, God, or whoever protects us all, smiled on me, because in twenty minutes a beaming Melanie came back onto the ward, gave me my credit card and forty-eight dollars, and told me she kept two dollars for herself. "I made change at the 7-Eleven."

"Thank you, Melanie," I said. She moved toward me and I thought she was going to hug me, but she stopped short, turned, and left the floor.

The story made the gossip grapevine of the hospital. While some people thought I was out of my mind, the most senior supervisors congratulated me on making a calculated, bold move. (They didn't know it was far from calculated but rather purely spontaneous.) "It worked," one said. "That's what counts."

In the early 1990s Melanie would be one of the first patients at MMHC to get on the new medication, Clozaril, which would revolutionize the treatment of chronic schizophrenia because it led to dramatic improvements never seen before. Decades later, she wrote to me at my request. She remembered my giving her my ATM card, and told me that my trust had meant a lot to her. She added the following account of her life, which she sent me so that I could include it in this memoir:

I was diagnosed with schizophrenia in 1964. I was sixteen. I spent decades and decades of my life wishing I had never been born.

Those were the dark ages of mental health care. Lobotomies had not yet been renamed psychosurgery. ECT was in a primitive state with being tied down to prevent broken bones and having a rubber mouthpiece in your mouth to keep from gnashing out one's teeth. I experienced shock treatments under the duress of threats to confine me indefinitely. For years afterward I had to make the most simple lists for each day. The list would say things like brush your teeth, get dressed, have cereal.

The hardest part of having schizophrenia is relationships. I looked OK . . . men were definitely interested . . . but my personality was completely undeveloped. I had no defenses or boundaries with men. I exacerbated these symptoms by using LSD, marijuana, and cocaine. These had a terrible effect on my ability to communicate either in writing or speaking.

I had no character or integrity to speak of . . . I just wasn't motivated to be a better person . . . I was in too much emotional pain to care. I didn't want to live. Twice I overdosed seriously enough to land in a hospital ICU. There were many other times I took all my meds. I found out that being poisoned hurts.

I discovered that cocaine could alleviate a condition of not experiencing pleasure.

There is a medical term for this condition. This was the result of dopamine inhibitors. Cocaine didn't so much as get me high . . . it just made me normal.

Pursuing drugs, overdoses, experiences of actually escaping from a strait jacket and also from leather four-point restraints I was, looking back, a good kid. I will never forgive myself for the pain I put my parents through.

At MMHC we had sing-alongs, poetry classes, and I sang loud and off-key but it inspired me.

I met my husband of 27 years in the locked unit at Mass Mental. I had never had a relationship that didn't cave in due to my diagnosis. Hank never abandoned me or judged me for being a deficient human being. After all, he too was diagnosed with schizophrenia.

We clung to one another as if in a hurricane. We were homeless and had many experiences of hardship like sleeping in cars when it was freezing out . . . other people were often kind to us. Eventually we got housing.

Thorazine had been a breakthrough enabling people to stay in their homes, helping to make strait jackets obsolete. The side effects of the "zene" medications were weight gain, tardive dyskinesia, loss of coordination and sluggish mental ability. I was on average a pretty and well co-ordinated young woman and the option of becoming like the Bride of Frankenstein on huge doses of Thorazine was impossible to accept.

I would become psychotic, go in the hospital, be given Thorazine (or later Stelazine, Prolixin, Haldol), and calmed down. This meant I would stop trying to set gas stations on fire or chasing women with baby strollers down the street while thinking their mothers were kidnapping their own little ones.

As soon as I was released I would go off the meds, lose the twenty to thirty pounds I had gained while on medications and have another successful six months to a year and then be back in the hospital, beginning the cycle over again.

In 1993 I was given Clozaril in a study conducted at MMHC and it worked as nothing before it had worked. My pleasure centers weren't as frozen as with dopamine inhibitors, my thoughts became fluid again and I was able to begin making a life for myself. I was for the first time able to be compliant with my treatment. I have had, thanks to this study, twenty-two good years. Now I want to give back.

I graduated from Johns Hopkins with an MA in the Writing Seminars in 1970. A big part of my life now is writing poetry . . . I have published three books . . . and continue to work and study and quite often am published in little magazines.

Hank and I somehow survived each other and have grown into a supportive and loving couple. This would never have

294 EDWARD M. HALLOWELL

happened without the support of my therapist who worked with me on establishing boundaries and pursuing my goal of investing emotionally in this relationship which most care-givers saw as a liability. I cannot imagine being happier in a true, equal relationship.

I spend quite a lot of my time volunteering at Mass Mental on the Human Rights Committee and also, working with others with lived experience of mental illness, serve on the Consumer Advisory / Commonwealth Research Center committees.

Every day is a full one and I am never bored. I was given Clozaril as part of a study conducted at MMHC, a study which gave me my life back.

All those years of intense suffering have left me a very grateful and appreciative and mostly happy individual.

I work on myself spiritually, morally and I believe I have achieved a fairly good outcome as far as character is concerned.

59.

One of the ways we learned our craft as residents was to present cases to senior clinicians, our supervisors, our teachers, in a group setting. One resident would present the story of one patient or family, and then the supervisor would interview the patient, with a discussion to follow.

These were not only valuable for our learning, but they gave the resident a chance to strut his stuff, or look foolish, or some of both. There was usually an audience of over thirty people, as all disciplines were invited.

One day I happened into such a conference just as it was beginning. The supervisor was Nick Avery, a brilliant man we all looked up to and upon whose every word we hung. The presenter was a resident named Peter Metz, a tall, trim, slightly balding man who had been at Harvard with me.

However, seeing Peter present now was my first significant exposure to him. One time in the coffee shop I'd mentioned a patient's blood urea nitrogen and creatinine levels to someone I was talking to, and Peter had leaned over and told me you don't really need to get both BUN and creatinine to assess renal function, that it was a waste of money. He was right, but it was a gratuitous, one-upmanship remark.

"OK, thanks," I'd said, hoping not to see him again anytime soon.

But now, as Peter talked, I became anxious. Although I hide it, I am a highly competitive person, and I could tell right away that Peter was not giving the usual patient presentation. He was presenting the real person, offering telling details that brought his patient to life. She was a massively obese woman who suffered from a chronic mental illness that was not quite schizophrenia, not quite bipolar, not quite depression, and not quite the consequence of a low IQ. She was a bit of all of these and more.

First-year residents were forever searching for what we called "therapy cases," since most patients under our care had chronic, severe mental illnesses. They were not amenable to insight-oriented psychotherapy, which is what we wanted to do. We didn't fully appreciate that we were at MMHC to get training in psychiatry, not psychoanalysis.

We were eager to treat people like us: well-educated, motivated, insightful, civilized, well-behaved, and neurotic. Neurotic did not have a precise meaning; generally, it meant not quite happy with life even though you probably should be, given your external circumstances. We wanted to learn how to talk with these people and help them become happy, which is exactly what we wanted for ourselves.

Peter had done what Semrad advised: He'd sat with his patient long and hard and come up with enough dynamic material to transform her from the inert lump other residents had treated her as to a woman who wrestled with conflicts we could all identify with.

It was a masterful job on his part, brilliant in fact, taking both immense imagination and smarts as well as hours of hard work and prolonged listening. I'd never seen anything like it. His work with this patient challenged all of us to find the "therapy case" in our patients.

I remember saying to myself, "Either I have to kill this guy or become his friend."

I chose the latter. Putting aside his remark about BUN and creatinine, I went up to him and congratulated him on his presentation. I quickly saw how deeply modest and self-effacing he was.

Even though we were different in style—he was reserved, I was flamboyant—we took to each other right away.

And then we stumbled onto an idea that changed our lives. "Do you play squash?" one of us, I can't remember which, asked the other. "Yes," came the reply.

We set up a weekly squash game. Since Peter and his wife, a pediatric cardiologist, moved an hour away to Worcester after he completed training, later on our game was biweekly. On the same afternoon, Peter would come in to play squash and have his piano lesson. (He is a virtuoso pianist.) After our game, we'd go out for a few beers and talk about cases, and our lives.

Peter and I have been playing squash and drinking beers on Tuesday afternoons for thirty-seven years. He is godfather to our daughter, and I am godfather to his. We are each other's best friend.

60.

Leston Havens let us residents use his office in the evenings to run group therapies. (We used to say, "Les is more.") Havens himself did not do groups. "I have a hard enough time keeping track of what's going on in the mind of one patient, let alone eight," he said. "But be my guest." It was typical of the esprit de corps of MMHC that a senior professor, an illustrious figure, would lend his office to us lowly residents to practice a treatment he didn't even think worked.

Since I was new to everything, I was new to group therapy, as was my co-leader, Jennifer Stevens. Actually, that is not quite true. I'd been learning about therapy for over a year since I'd started seeing Dr. Khantzian as a patient in psychoanalysis during internship. And I was learning about group therapy in the group run by Max Day that we were all required to participate in.

What I was new to was the practice of doing therapy myself, group or otherwise. We residents were encouraged to ask someone from a different discipline to co-lead groups with us, so I had asked Jennifer, who was an OT, or occupational therapist, on the inpatient unit. Because I liked her, it almost felt like asking her out on a date.

The supervisor of the group program, Bessel van der Kolk, who would go on to become a world authority on trauma, had assigned Dr. Allan Hobson to be our supervisor. A professor at Harvard Medical School, Allan was famous in his own right, doing pioneering research into sleep while also trying his best to debunk Freud's theories about dreams. Not just a biological psychiatrist, as I would learn, Allan was also a brilliant and intuitive therapist. Jennifer and I met with him briefly before our first group meeting. He just told us to jump in and see what happened. My kind of guy!

We had eight outpatients in the group, which had been running for years before we inherited it from another pair of group leaders who'd moved on to the next step in their training. It was yet another example of how patients helped us out, putting up with losing the group leaders they'd gotten accustomed to and getting to know new leaders just so we could get training. What the patients got in return was a low fee, but I often wondered if we shouldn't pay them.

Despite Havens's skepticism, running a group was a requirement in our training. Some of the group supervisors could be pretty dogmatic, if not ridiculous. For example, one day I was walking down a corridor of the hospital when I passed an open office door. Pausing, I looked in.

There was Dr. Joyce Baker, a fellow first-year resident, sitting by herself in a straight chair in a circle of eight other straight chairs, all of them empty.

"What are you doing?" I asked.

"Shhh. I'm not supposed to talk to anyone except the members of the group."

"But no one's in the group. No one's here."

"It doesn't matter," she whispered, as if the ghost of her supervisor might hear her. "*I'm* here. And I'm supposed to sit here for the full hour and fifteen minutes, whether someone else comes or not. My supervisor says it's still a group, even if I run it alone. I'm supposed to sit and feel what it's like."

"Feel what it's like to run a group therapy with no one in the group?"

"Yes! Now go away."

"Joyce, would you like me to join your group?" I volunteered. "I'd be glad to be your patient in group therapy. Then you could tell your supervisor you had at least one member."

"Ned, you're bad. I can't do that and you know it. I'm not even supposed to be having this conversation."

"Like this is a forbidden interaction? Forbidden interactions should be juicier than this!"

"Stop it!" Joyce said in a loud whisper.

"Joyce, this is absurd. You've gotta agree. It's like a scene out of an Ionesco play."

"What kind of play?" Joyce asked.

"Ionesco. He's a French playwright who writes absurd plays, and this is absurd."

"I know, but I have to do it the way he wants me to. He's insistent."

"Who's insistent?"

"My supervisor," Joyce whispered, as if the empty chairs had ears, "and he's really strict. I don't want to get in trouble. I have to do what he asks me to do."

"Even if it makes no sense?"

"We're here to learn, right? Maybe there's something to it. *Please* go away."

"OK, I'm sorry. Didn't mean to interrupt—" I almost said "interrupt your group," but stopped myself because I didn't want to sound sarcastic. Joyce was what my family would have called a really good egg, so I didn't want to annoy her more than I already had.

Being the good student she was, Joyce told her supervisor about my interrupting "the group," and her supervisor took her to task for talking to me, telling her that she had "let the group down" by allowing me to violate the boundaries of the group and allowing me to "upset the group's integrity." It seems I had denied Joyce the opportunity to feel "the full existential tension," as the supervisor put it, of running a group with no one in it but the leader.

After I apologized to Joyce, she laughed it off. She had a mature approach. "I never know what to expect. I just try to take the best of what each supervisor has to offer and then make up my own mind later on."

Thank God, our supervisor had given us no such rigid rules by which to run our group. "Just jump in." The more formal side of me wondered if we should make interpretations. Interpretations were the bread and butter of therapy.

"Interpretation" is a word out of psychoanalysis. A therapist interprets the foreign language of unconscious material, group dynamics, connections between seemingly disconnected events, the meaning hidden in what appears to lack meaning, or the emotion buried in a blank stare. Making interpretations is like detective work, gathering disparate clues to find the cause or significance in a certain statement, behavior, dream, action, or event. Making interpretations is a lot like what I did as an English major while analyzing a poem or a novel, which is one reason majoring in English helped me become a good psychiatrist. Same with being a writer. Professor Alfred had told me becoming a doctor would help me as a writer. As he put it, "Writers and doctors are always on the lookout for the telling detail."

We got zero training in making interpretations in medical school or internship, so college or residency was the time to learn how to do it. But even in residency, we were more or less on our own, learning from whatever books and articles we chose to read, and from our various supervisors.

For example, after starting the group with Jennifer, an obvious interpretation we could make might be that if a member arrived late, that meant they were avoiding the group because the old leaders had left. As far as psychodynamic thinking went, you were never just late—or early, for that matter. Every action had its cause. The time you arrived was a reflection of an unconscious, or maybe conscious, motive.

We didn't learn this stuff from a textbook, although there were many such books for us to peruse. The psychologists got the book learning. Psychiatrists sat with patients. We also learned from case conferences, from seminars in which we read papers from academic journals, and from supervision sessions, as well as from talking with each other. I love this way of learning, and I loved being a resident. To this day I am still learning in the same ways, even though I long ago completed my formal training.

There were a host of axioms and unwritten rules we absorbed pretty quickly. "Every gift is a bribe," which meant we should not accept gifts from patients, nor should we offer gifts. "No PC," which meant no personal contact, or touching of patients. Shaking hands was OK, but that was it. "Every relationship is ambivalent," which meant people never love or hate anyone a hundred percent.

And then there were the many Semrad sayings we heard from supervisors. Some of my favorites: "Is there any other way to learn than the hard way?" . . . "If you have somebody to be mad at, then there's a place for all the energy to go." . . . "What one feels is a part of life not amenable to reason." . . . "Sorrow is the vitamin of growth." . . . "Pretending that it *can* be when it *can't* is how people break their hearts." . . . "It *hurts* to think straight." . . . "You can have definite opinions only when you don't know anything about a subject." . . . "This is one of the eternal questions: How much are you going to pay for what you get?" Susan Rako and Harvey Mazer, two of his former students, collected some of his quotes in a book called *Semrad: The Heart of a Therapist*. A slim volume, it is the only written record we have of Semrad.

Havens's office, which used to be Semrad's office, had ten chairs of various shapes and sizes, all upholstered and much more comfortable than the straight chairs in Joyce's group room. There was an attractive, albeit worn, oriental rug on the floor, bookcases along one wall, and Havens's desk at the far end of the office. There were a couple of small coffee-table-type furnishings, some standing lamps so we didn't have to use the antiseptic overhead light, and three large windows (with much larger panes than those in my office) looking out onto Fenwood Road in the dusk of early evening in Boston in October.

The group was set to start at six o'clock, but Jennifer and I arrived early, about 5:45, and made a decision to sit opposite each other rather than side by side. "I don't want them to feel like we're king and queen," Jennifer said, and I agreed.

We waited. I thought of Joyce. "What if nobody comes?" I asked. "What if they're so pissed that they got handed off to us that they boycott?"

"You think they will?" Jennifer asked.

She was beautiful, with long auburn hair and a slightly freckled face. I was very drawn to her, and we even dated for a while after the group ended, but it was not to be. She had her eye on another guy.

By six, no one had arrived. I felt like a host when no one comes to the party. Maybe they're fashionably late, I thought to myself. Five past six, still no one.

"Are we in the right room?" Jennifer asked.

"We have to be. This is Havens's office, and this group has been meeting here since before the last leaders took over, like forever."

At that, as if beckoned by the ghosts of groups past, three members arrived and took the seats they must have been taking before, as there was no hesitation in their choice of where to sit. Over the next five minutes, the other five members arrived, causing me no end of relief.

The group was kind to Jennifer and me. They introduced themselves and thanked us for picking up where the old leaders had left off. "Don't worry," one named Maxwell, a thin bearded man, said. "We'll give you a nice honeymoon before we start to chew you up."

"Shut up, Maxwell," another said. He was named Bernie, and looked like an ex-football player. "You're gonna scare them off."

"We're tame," said a third member, Francie. She looked as if she had once been utterly beautiful, a bit like Lauren Bacall, but the meds, cigarettes, booze, and years had taken their toll. It's a look I came to recognize easily and see often, a face's pentimento. "We're chronically mentally ill," Francie continued. "What could be tamer than that? No matter what we do, we don't know what we're doing and you have the right to lock us up any time if you want to."

"That's not correct," George piped up. "We have certain unalienable rights." He said this with great mock pride, clearly ironic. George had the red nose of a drinker, but also dressed very well.

Despina was supportive. "George, you know you always undercut yourself. We *do* have certain unalienable rights. How are we ever going make progress if we can't take ourselves seriously?"

"You still think you're Greek nobility, so it's easy for you to feel pride," George replied. "What am I but a chronic drunk with a brutal past and no future?"

Jennifer and I sat a little dumbfounded as the group took off. I wondered if they did this every week. What did they need us for? Once again, as usual at MMHC, the clinicians needed the patients more than the patients needed the clinicians. I was smart enough not to interrupt the process, and thankful that Jennifer was as well.

"Have you had a bad week, George?" asked Despina. Despina did indeed look like royalty, Greek or otherwise, dressed in a flowing blue gown and bedecked with paste jewelry.

"You could say that. I got laid off. So what else is new? But my boss was a real asshole so I'm just as glad."

"Fuck him!" Alice jumped in. "It's not right they just fire you like that. What reason did he give?"

"Thanks," George said, "but I had it coming. I'm always late. Why does this world make such a big deal about being on time? I hope these leaders aren't like that."

Seeing an opening, I asked, "Were the leaders before us sticklers about time?"

"Oh, yes," George said. " 'We start on time. We don't want to reward lateness,' they would say. What they didn't know is that we knew that that line came from Dr. Gutheil, their supervisor, so when they'd say it, we'd say 'Thanks, Dr. Gutheil.' They'd get embarrassed and not admit they were stealing his line, but we knew, and they knew we knew."

"How *did* you know?" Jennifer asked.

"You're very pretty," George said. "Much prettier than the lady you replaced."

"Don't talk like that," said Lucas. "It's not respectful. Remember, we all agreed that respect is the main rule of the group." Lucas, who had once been a schoolteacher, still talked and looked the part.

"OK. So are you now the policeman?"

"We're all the police," Lucas said. "That's what we're trying to learn here, how to police ourselves. I wasn't putting you down, just reminding you that you agreed, with the rest of us, that being respectful was a good rule to stick to."

"Why isn't it respectful to tell her she's pretty?" George asked.

"It's not exactly what you say right off the bat to a new leader, but the really disrespectful part was your dissing Dr. Kathy."

"I didn't like Dr. Kathy. I never got to tell her that."

"What held you back from telling her you didn't like her when she was leading the group?" Jennifer asked.

"I guess I felt intimidated," George said. "Let's face it, you guys, the group leaders, the docs, all the people in charge here, you're on a totally different status level than the patients. In your eyes, we're inferior beings. In the eyes of the world, we're inferior beings."

"That's not right," Alice said sharply. "We are *not* inferior. If we can't stand up for ourselves, who will?"

"That's why we're here, to help each other do that, out in the world," Lucas said. "It begins here, then we take it outside. But we gotta respect each other in here."

Francie started to cry, and Despina offered her a box of Kleenex— Kleenex was always in great demand at MMHC.

"What's wrong, Francie?" Alice asked.

"It's just so hard," Francie said. "Everything you say is so right, Alice, but what George said is right, too. The world really is against us. It is *so hard* out there. I don't know why I even get up in the morning. I get a little bit of hope and try to make myself look presentable, even look pretty, like I used to be, you know, like I don't have a mental illness, but I just can't do it. I don't know how to do it anymore, I lost that somewhere, I used to be a normal person, a normal woman, a pretty woman, if you can believe that. And when I get up, if I don't think twice, I can actually imagine I am that person that I used to be. But then I start to try and make myself look like her and I just don't know how to anymore. That face is gone, even though I can almost see it in the mirror. But I know when I finish dressing and putting on makeup, I know the world can still tell I'm a crazy. I walk out the door and say to myself, 'Hold your head up high, Francie, just like Mom told you to do, just like *she* always did no matter what.' And she'd tell me, 'Always remember you have dignity, and no one can take that from you except you.' I give myself this pep talk as I walk out the door, and then the first person who walks past me gives me that horrible, horrible second look, and I feel it like a slap in the face, all my dignity goes to hell and I know I'm just a piece of garbage, like where do you get off, Francie, you're just a total nobody, a piece of garbage. Living like this, it's just *so hard*."

"Oh, shut up, Francie," said Maxwell. "Stop being a crybaby."

"She's not being a crybaby," George spoke up. "She's speaking for every single one of us in this room, including you, Maxwell. Maybe not the leaders, but all of the rest of us feel that way."

"Don't think we don't have our moments of feeling pretty shitty," I said.

"But, Doc," George said, "look at you, and look at Lady Doc over there—"

"I'm not a doctor," Jennifer corrected. "I'm an occupational therapist."

"Whatever," George said, "you're on one level, you and Doc here, and us in the group, we're on the garbage heap, like Francie said."

"I am *not* on a garbage heap!" Despina stated indignantly.

"Don't you tell me what to say," Maxwell interjected.

"Oh, stop it, Maxwell," George replied. "We all know you can't stand to be sad, so you try to turn sad moments into fights. You've been with us long enough, we're wise to your tricks."

"Fuck you."

"Back atcha," George replied with a smile.

"You all know each other pretty well," I said.

"We do, we do," George said. "But we don't know you and Princess Grace over there."

"You'll get to know us," I said. "But don't you think we should stick with what Francie was saying? All of you live out in the world, and you've learned how to live with what you have, but it's hard."

George responded. "Thanks for the support, Doc. But you couldn't possibly know what it's like to be one of us. Have you ever suffered from, let's say, manic depressive illness? Or had anyone close to you come down with that? Have you seen your family ruined, like Despina has? She used to be like a queen in Greece, then it fell apart because her family went nuts. I know you're trying to help us, Doc, but how can you possibly understand what Francie feels when she's getting dressed in the morning to go out into the world?"

61.

Havens looked like Jack Paar, the first host of *The Tonight Show*. Handsome, trim, an avid tennis player (I don't know why, but many psychoanalysts love tennis). He was also at heart a humanist. He quoted literature all the time. Joseph Conrad was one of his favorites.

I never got to know him well. Not many people did. He allowed me to develop a personal relationship with him at the distance he felt comfortable with. My hunch is that he sensed how needy I was (am), and so he kept a distance lest he be asked to give more than he wanted to or could.

Back when I was a first-year resident, one of Les's sons died. Even though I didn't know Les well, I'd read his classic book, *Approaches to the Mind*, and already revered him, so I sent a note of sympathy. To my surprise, he wrote back a kind note of gratitude, even though we'd never met.

You could say that note sealed the deal for me. I felt a bond with Les ever after. I was hardly alone. Many people felt they had a special bond with Les even though they spent almost no time with him one on one. He noticed the little things. Walking through the corridors of MMHC, for example, when he and I would pass each other, he always said, "There he is."

There he is. How to take a puny little resident and make him feel like he's on top of the world. He did this so naturally and with so many people. Was it manipulation? No, I don't think so. It was empathy. Les *knew* we residents were starved for recognition. In three simple words he gave me that, many times.

He liked to supervise residents in groups rather than individually, and he often talked with us about the history of MMHC and the field. "The Viennese took over psychoanalysis in Boston faster than the British dispersed the Indians. The embodiment of all they stood for you could see in Grete Bibring. She was a brilliant woman, a graduate of Harvard Medical School and a top student there. I knew her well because she supervised one of my cases at the Institute for four years. She and her tribe—the classical Viennese analysts—held Boston in a stranglehold for years—feeling they had to protect their leader, Sigmund Freud, at all costs."

Havens had an animated, engaging way of speaking, even to a small group like the three of us he supervised. "They were right to do so, no doubt, because without their strict authority cowing everyone into submission, people might notice the great king was missing some clothes.

"The kind of analysis and training Grete Bibring insisted upon could be summed up in a story she told on herself," Havens said to us. "She recounted that one day her little daughter said, 'Mommy, my friend says her mother is like a soft, fluffy doll.' But then Bibring's daughter pointed to a porcelain figurine standing on a table across the room, beautifully painted but hard and metallic. 'Mommy, you're more like that.' 'And it's true,' Bibring said. 'I am more like that. That is how an analyst ought to be.'"

I asked, "Grete Bibring actually told that story on herself? Like she was proud of it?"

"Yes," Les said. "The Austrian contingent felt a deep loyalty to their leader, and what they perceived to be the method Freud recommended. If you wanted to get ahead in this town, you had to go to the Institute and do it their way. The development of emotion in conducting a psychoanalysis took a back seat to clarification, interpretation, and intellectualization. It was all about being smart."

Les stretched his arms, rubbed his shoulders, put his feet up on his desk, and leaned back in his swivel chair. "And then we had Elvin," he said with a smile. "Elvin put emotion first. He would say, 'If you have to tell somebody something, it's already too late.'

"His famous formulation was that you acknowledge, bear, and put into perspective the patient's feelings and experiences. But notice that the first two, by far the most important, draw not on intellect but on emotion. People in the Institute mocked him. But he didn't care, or at least he said he didn't."

"It had to hurt his feelings," I said.

"It probably did," Les replied, "but people also loved him. He didn't care about that, either. He just wanted to teach and impart what he knew about helping people. And he knew a lot. He just didn't play their game. He didn't have the turf to protect that the Viennese did. He didn't like to fight, so he'd go to meetings and allow all the smarty-pantses to win. But he won the battle for what matters most."

Havens went on, "This is why innovation in our field is so difficult. Everyone is protecting turf and attacking anyone who proposes anything new. Heinz Kohut is a good example. When he wrote *The Restoration of the Self*, which stressed empathy in treating narcissism, the classical analysts went ballistic. If you ever want to read a truly savage review, a cruel and completely gratuitous attack, read Harold Blum's review of that book in the *Journal of the American Psychoanalytic Association*. Kohut was old, and I truly think that review is what killed him."

I could see from what Les told us that my analyst, Khantzian, was more Kohutian, more Semradian, than Viennese. That made sense, since he had trained under Semrad.

I would also learn, mostly from Les, that what I had hoped to do at MMHC—explore the deep recesses of the human mind—was more the purview of a psychoanalytic institute than a state mental hospital. We were learning psychiatry, not psychoanalysis.

I was at a juncture as a second-year resident. If I wanted to pursue psychoanalytic training, now was the time to start. Most of us harbored the ambition of getting analytic training—it seemed to be the gold standard if you wanted to be a deep student of human nature.

But even if I had been able to afford the time and money (I couldn't), I had to look at myself and say, You're not analyst material. You're too impatient. You can't sit still and listen that long. You can't *not* say what you want to say long enough for the patient to ramble on, freely associating.

I had to look no further than the annual MMHC dinner for evidence. At the end of each year, the hospital held a dinner at the Harvard Faculty Club to celebrate the year and bring us all together. Each year there was a guest speaker, some luminary in the world of academic psychiatry. I looked forward to these lectures, as they were usually excellent.

One year, however, the speaker, despite his international renown, was awful. He was mailing it in, with no end in sight. I literally could not sit still for it. As I happened to be sitting next to a window, without giving it a second thought, I opened the window and jumped out. Fortunately, the banquet room was on the first floor, so I landed safely with no injury and walked away, relieved to have escaped intolerable boredom.

It was also fortunate that the table I was sitting at was in a corner, out of sight from the big shots at the head table. A psychoanalyst cannot jump out the window if his patient bores him, and every patient in analysis must at one time or another bore his analyst. But I had to be honest. Even though I yearned to be a psychoanalyst, I would have hated the actual job.

Dr. K. agreed with me, although he didn't come right out and say it. Instead, he said, "It's not for everyone."

All the same, I would read the description of analytic training and think, *What could be better than that?* But it's a journey I will never take.

How much they must know, these modern-day Yodas. No longer in the stranglehold of the Viennese, I like to believe they are cut much more out of the cloth of Kohut, Havens, and Semrad now than ever before.

Les Havens, one of the greatest teachers to ever come down the pike—not only in psychiatry but on the subject of human nature in general—died two days before his eighty-seventh birthday in 2011.

62.

Karen sat in a chair in my office picking at her clothes, looking at the wall, the floor, here and there, like a bird on a wire. She was dressed in layers: gray shirt covered by red shirt covered by plaid jacket; jeans covered by green tattered skirt; bright white sneakers she must have just found in the clothes depository; gray wool socks; various gold-colored chains around her neck; a knitted rainbow-colored beret-like hat over her caked-down black hair. She had a torn paper shopping bag full of this and that at her side.

One of the "experienced" patients at MMHC, Karen had been teaching residents for years. Now in her forties, she'd been assigned to me. In years to come, she'd leave the hospital thanks to the miracle drug Clozapine. But in 1979 we didn't have that yet in the United States. (It was available in Europe, but the wheels of the FDA grind slowly.)

Karen was quietly psychotic. My job was to forge a relationship with her. I was to do this by using that time-honored method Dr. Semrad championed: sitting with her.

Karen couldn't communicate, at least not with words. I had to intuit what was on her mind. In addition to sitting with her, Karen's therapy included prescribing antipsychotic medication, which did help somewhat. I also decided what privileges she could have, i.e., how much freedom she got to roam the hospital, even to leave the hospital if she seemed able to handle the responsibility.

I was learning the value of sitting, the value of connecting.

I learned from Karen what it feels like inside *me* when I cannot converse with another person, even though she speaks English. It feels frustrating, aggravating, and sad. Semrad would call this "the empathic diagnosis." By noticing what I am feeling, I may be getting in touch with what the patient is feeling. Or I may not. There's no proving it one way or the other.

But I was taught—and I obeyed—not to do what I most wanted to do: terminate the session because it was pointless. That would be denying the value of the patient's time and reality. Wasn't Karen as entitled to my attention as a patient who could converse meaningfully with me?

I kept sitting. I would watch Karen, listen for whatever she might say, cross and uncross my legs . . . and gradually come to see the point in it. Karen would mutter phrases, many taken from old advertising jingles, that were perhaps like rosary beads to her; phrases like, "Ivory soap floats," "M&Ms melt in your mouth, not in your hand," or "You're in good hands with Allstate."

I'd try to play off one of those and see if I could strike a chord. For example, one day when she muttered, "You're in good hands with Allstate," I said, "I bet you really hope you're in good hands here."

She made no reply and went on, leaving me to wonder what she did with what she heard. That's the mystery of the psychotic mind. Most people would say she did nothing with it, that my words meant nothing and had no impact.

Weeks later, I would learn differently in the Community Meeting, the biweekly meeting of our unit's patients and staff, which was run along the lines of a Quaker meeting.

Karen, out of nowhere, spoke up and proclaimed, "We're in good hands here!"

People's jaws dropped. Professor Stein, the ex officio patient leader, said, "Karen, it's so good to hear from you!"

Karen didn't respond but just kept looking around like a bird, as she always did. Still, it was a stunning moment, at least for me, and, I could tell, for Professor Stein.

I met with Karen weekly. She regularly earned privileges to leave the hospital on group walk—which she enjoyed, it's when she could buy her cigarettes—as well as earned privileges to go to a halfway house for overnights.

At that, she balked. No overnights for Karen. She'd developed what I learned was called an "institutional transference." Nonpsychotic people do this when they fall in love with their alma mater, for example, or, like me, with the team they root for, or the city they live in; institutions that have no feelings get treated as if they do have feelings, as if they were people who cared in return.

In Karen's case, the transference was with MMHC. When she gained privileges to sleep outside the hospital, she'd eschew the halfway house and instead dig a hole up against the cement foundation of MMHC, curl up, and sleep in that hole. The police had grown accustomed enough to it that they let her sleep there unless the weather was too cold or wet, in which case they'd call us to bring her inside.

One night I was the one on duty when the call came in to go outside and get Karen. The sight of her curled up in the hole she'd dug for herself, wrapped in her many layers of motley attire, etched an image in my mind I've carried ever since. Talk about a safe place; even though it was against an inanimate slab of concrete, Karen had found hers. She was sleeping as peacefully as an old hearth dog, in good hands.

63.

Early in my first year at MMHC I attended a Grand Rounds given by one James T. Hilliard. All I knew was that he was an attorney, and we were required to attend his presentation on risk management.

I went in thinking this was going to be a total waste of time. I didn't need to be bothered with laws, did I? Just do your job the best you can, treat patients with care and respect, and you'll be fine, right?

In less than an hour I was reduced to a trembling pool of petrified protoplasm. Jim, a low-key, obviously smart and experienced attorney who knew his stuff cold, began by saying to us newbie psychiatrists, "It's not too late. You can find another career and never worry again about what I'm going to tell you. The fact is, if I were you, I *would* get out now. I couldn't stand living with the risk you all take on every day just by walking into the hospital. Then, when you see patients and start signing your name, the risk goes through the roof."

This guy was not putting on a performance. He was telling us about life as he lived it every day, defending doctors who were being sued, showing us what we didn't see since we were just clinicians, not lawyers. He was urging us to wise up, to stop pretending that everything works out for the best as long as you are a good boy or girl, and if we were not taking his advice and leaving the practice of medicine altogether, then at least we should take *very* seriously the dangerous spots we put ourselves in every day. I learned terms I'd never heard before, like "deep pockets."

"That's a legal term?" I asked.

"Yes," Jim said. "People who sue people want to sue people who have deep pockets, or who have insurance that has deep pockets. You understand what I mean? People want to get a big judgment so they can retire because of your mistake."

He went on to tell us about the dangers of insufficient documentation ("If it isn't documented in the chart, it never happened"), failure to communicate with patients ("Doctors who don't take the time to listen to a patient's complaint are the doctors who get sued"), flying solo ("When you start believing you know more than everyone else, that's when you get in trouble"), having sex with a patient ("Don't do it, don't do it, don't do it"), forgetting to ask about suicidal thoughts ("If a depressed patient commits suicide and you haven't documented that you asked about suicidal thoughts, you are exposed to a wrongful death suit").

Jim was so convincing that I wanted to start carrying a tape recorder with me everywhere I went. I wanted to document every interaction I had, every thought that crossed my mind, every word and action from every patient. In that hour I learned to fear my job, fear patients, fear the legal system, fear attorneys, and most of all, fear my own fallibility.

I'd be much safer, I thought, if I quit then and there and applied to law school. An attorney who also has an M.D. is a hot commodity. I could make a lot of money defending doctors, maybe lobbying to get the laws changed, finding ways to tackle the injustice of it all, while protecting myself from the seeming inevitability of getting sued ("It's not *if* you get sued," Jim told us, "it's *when* you get sued").

It was a short-lived reverie. My lot was cast. Still, after hearing Jim Hilliard's talk, I never felt the same as a doctor. An element of fear seeded my system that some would say was good—Bill Clinton once commented that suing doctors is good "because it keeps them on their toes"—but I knew for a fact it was bad. Doctors fearing patients is as harmful for medicine as patients fearing doctors.

But from that day forward, I had to learn to live with it. We all did.

64.

I hope I'm not being too cynical, but—" I said to Dr. Taube, my supervisor on outpatient cases.

"You cannot be *too* cynical," Dr. Taube interrupted. "Your problem is you're not cynical *enough*. But I can help you with that," he said with a wry smile. "I am a *master* cynic."

Irv Taube was a shortish man, always impeccably dressed in a suit, usually Brooks Brothers wool or gabardine, with beautifully contrasting shirt and tie, who carried an elegant, brown (not black) leather doctor's bag with him everywhere. He had trained at MMHC back in the heyday of Semrad but didn't glorify those years. In fact, he didn't glorify much of anything. His view of life was dour at best. And yet he loved teaching.

Irv was short for Irvin, not Irving. Like his doctor's bag and his name, Irv did not fit any mold. He appeared vain but wasn't. He appeared soft-spoken but was highly opinionated. He appeared slight but was strong as an ox. He often appeared to be distracted but paid rapt attention. He claimed to know little and have few answers, but he knew a lot and had buckets of answers. I looked forward to every meeting with Irv just for the experience of being with Irv.

I had started my fellowship in child psychiatry, but we were given one supervisor for the adult cases we had, and Irv was mine. We met at "the lunch hour," as he called it. After I arrived, he would carefully extract his meal from his bag. "I hope you don't mind if I eat? It *is* the lunch hour," he'd say. "Please feel free to bring your lunch next time." And I would always reply, "Of course, please do eat. Don't worry about me, I usually skip lunch." It would be unthinkable for him to pull out lunch while meeting with someone who was not eating lunch without first asking that person's indulgence.

Once in a while, he would caution me, "You shouldn't skip meals. One of the bad habits doctors get into is not taking care of themselves. They get so wrapped up in taking care of their patients that they neglect their own needs. And you know what that leads to, don't you?"

"Bad stuff," I said.

Irv smiled and nodded, but ever the Socratic teacher he then asked, "Could you be a bit more specific?"

"Burnout."

"Yes, but what else, more common than burnout? Most doctors are too driven to burn out. Instead they just keep dragging along, dreading every day, but showing up just the same. But what else do they do when they neglect their own needs?"

"They get them met in dangerous ways, like drinking too much, taking drugs, going in on harebrained financial deals, having sex with the wrong people?"

"Yes, exactly," Irv said, carefully wiping his mouth with a napkin. "If you don't get it where you should, you get it where you shouldn't. That's my long version of telling you not to skip meals. Why do you skip lunch, by the way?"

"I want to lose weight," I admitted.

"I thought you were going to say because you're too busy. I'm glad you didn't play *that* card. You want to lose weight? Why? You look like you're in good shape. Do you get exercise?"

"Yes. I play squash with Peter Metz every week."

"Oh, good. You know I supervise Peter as well? He is a very smart man."

"He's my best friend," I said proudly.

"Well, you have good taste in friends, then. He is not only smart, but he is a good person."

"I agree," I said. "I will tell him you said so." I always tried to relay compliments, as MMHC teemed with backbiting, nasty gossip.

"Before we talk about your case, I would be remiss if I didn't tell you that skipping meals is no way to lose weight. You have to change your eating habits. But you don't need me to tell you that, do you?"

"You're reminding me of what I know. I'll do my best."

"Good," Irv said. "So what's going on with Mr. Sloan?"

I'd been presenting a Northeastern senior to Irv for the past couple of months. He was twenty-two years old and he'd come to the clinic because he was struggling academically. He'd seen a therapist over the summer at his home in Virginia who'd diagnosed him with depression and put him on an antidepressant. I had continued the medication but was trying to go deeper into his issues. "He seems to be doing all right, but he is still struggling. He works really hard and should be doing better."

"What does he want from you?" Irv asked.

"He wants relief. He wants to do better with less work. He wants to be happier," I replied.

After taking another bite of his sandwich, which appeared to be chicken with lettuce and mayo, Irv asked, "And what do you want from him?"

I was taken aback. No supervisor had ever asked me that question. "I don't know. I've never thought about what I want from him."

"Really?" Irv said. "Actually, I am not surprised. It's not common for a supervisor to ask a trainee what he wants from a patient, is it?"

"No, it certainly isn't."

"But why not?" Irv asked, rhetorically. "This is life, after all. When a patient meets a doctor, and a doctor meets a patient, the larger rules of life still apply. I know you like literature. Do you remember the line in *Heartbreak House*, 'Do you think the laws of God will be suspended in favor of England because you were born in it?' Well, I sometimes think around here people think the laws of human nature will be suspended in favor of psychotherapy because we practice it. In *every* human interaction between two people, each person wants something. Never forget that. It is a fundamental law of life."

"That sounds so cynical," I said.

"I told you already I'd teach you to be more cynical. Honestly, Ned, I am just trying to help you get over your naïve assessment of human nature. It's wishful thinking. I hate to tell you, but it is. And it will get you into trouble, if it hasn't already."

"I'm sure you're right."

"So let's come back to your patient. Think for a minute, what do you want from him?"

"I want him to get better," I said, stating the obvious.

"What else?"

"I want him to get what he wants."

"Really? What if he wanted to kill you? Would you still want him to get what he wants?" Irv folded his napkin after finishing his sandwich.

I laughed. "Well, of course not."

"Don't laugh. A lot of our patients would like to kill us. The psychotic ones tell us so, which is why we can learn so much from them. The outpatients are much more artful at concealing the fact, even from themselves."

"I really do not think that Mr. Sloan wants to kill me."

"I agree with you. I don't think he does, either. But since I promised to school you in cynicism, I just want you to be aware that there is murder in everyone's heart, only most people have no clue it's there. Most people are blissfully unaware of who they truly are, and they positively, absolutely, one hundred percent do not want to know the truth, which is probably all to the good. They'd be shocked if they ever saw themselves as they truly are. That's why offering people the truth is such a perilous undertaking. It's what makes being a psychiatrist dangerous, at least being a good one. People would rather kill than face up to who they are."

"C'mon, Irv," I said, "I really don't think my patients want to kill me."

"I know you don't, and you are right about that. I just want you to know they have it in them, and even more importantly, I want you to know that *you* have it in *you*."

I flashed back to how I attacked Uncle Unger's hats. Irv could see that something had clicked. "That struck a chord?"

"Yes, yes it did."

"So we now agree?"

"I guess so," I said, surprised, if not shaken.

"Good. Then let's come back to what you want from this patient. You want him to get better, you want him to get what he wants, as long as it's not to kill you or hurt you, I assume. What else?"

Thinking of Uncle Unger's hats had lowered my defenses. "I want him to like me."

"Bravo!" Irv said. "Good for you! *Exactement*! That's the really dangerous one. Especially for you. If you don't mind my telling you, you are a real people-pleaser, which is fine, and much better than its opposite, except when you are doing psychotherapy. If all you want to do is make the patient like you and keep him happy, you will never allow him to feel enough pain to want to change and grow."

"Yes," I said. "You're right about me. I'm actually working on that. I can be meaner than you think."

"It's not about being mean," Irv said.

"I know, I know. I was just saying I'm not as much of a people-pleaser as I might seem. If I could go back to Mr. Sloan?"

"We never left him," Irv said with a smile. "This work is all of a piece." At that point, Irv pulled his knitting out of his doctor's bag. He often knitted during supervision. "It helps me concentrate," he'd explained when pulling his yarn and needles out the first time.

"His troubles might not just be because of depression," I suggested. "I'm just learning about this thing called ADD over on the Children's Unit. Do you know about it?"

"Why, do you think Mr. Sloan might have it? I thought it was just something hyper little boys had."

"That's the stereotype, but adults can have it, too. Do you know Paul Wender?"

"Yes!" Irv said enthusiastically. "He was a resident here. We were about the same age. He was one of the first residents to buck the tide of psychoanalysis back in the early sixties, which was the heyday for analysts. It took courage not to kowtow. But Paul didn't kowtow to anyone. Used to drive his analytic supervisors nuts. But he was smart enough to get away with it. So is he writing about ADD?"

"Yes, he's one of the people who's saying it continues into adulthood, or that it *can* continue into adulthood," I said.

"And you think Mr. Sloan might have it?"

"I was wondering about it. Of course, I'm just learning about it, so I am starting to see it everywhere. I actually think I have it myself!"

"But how could you? You've put up a stellar academic record. Doesn't ADD mean you struggle in school?"

I loved how open-minded he was. This was what made MMHC so great. During my training, it was open season on truth. We could raise any idea, and our teachers would take it seriously. It was a great time to train, because everyone knew how awfully much we didn't know.

Not to say there weren't factions, but they weren't vicious factions, as in the old days of Grete Bibring and the ominous Institute. The factions of our day got along, respected each other, and let us residents feast on the smorgasbord of their disagreements. You had the flamboyant Allan Hobson, one of only a few full professors at MMHC, a sleep researcher as well as an MMHC-trained general psychiatrist dosed heavily with Semrad, trying to take down Freud, especially his theories of dreams.

And then we had Joe Schildkraut, another full professor, father of the catecholamine theory of depression. In 1965 he wrote a classic paper published in the *American Journal of Psychiatry*, called the Green Journal because—guess why?—it had a green cover. In what became the journal's most cited paper, Joe presented his findings that catecholamine dysfunction (especially norepinephrine) was strongly associated with major depression. That paper changed the direction of psychiatry, supporting the development of antidepressant medication and thus allowing the field to break free from the orthodoxies of strict Freudian psychoanalysis toward a broader, more eclectic approach, embodied in teachers such as Les Havens ("Try whatever you think will work") and Doris Benaron ("Let the patient lead you"). Like Hobson, Schildkraut embraced psychodynamic thinking. Both men saw the issue not as *either* psychodynamic *or* biologic, but *both/and*.

And then you had the fast-talking pepperpot with the genius IQ Tom Gutheil, son of the famous psychoanalyst Emil Gutheil, holding up the best of the dynamic psychoanalytic approach. When one of my patients pulled out his own teeth after his mother's suicide, ripped out the sinks and toilets, and flooded the inpatient ward, Tom suggested it was a manifestation of oral rage. "The porcelain of the sinks and toilets represented the enamel of his teeth, which he felt a rage to remove."

To balance that, we had the ever-practical Bill Beuscher, who had been the faculty member in charge of the unit when I was a first-year resident. Bill was a psychiatrist's psychiatrist. As I mentioned before, he just wanted to "cure people up." Always dressed in a seersucker suit, he'd sit during rounds rolling his tie up and down as he listened. He seemed to be lost in thought, but he never missed a beat.

When I told him of Tom Gutheil's interpretation of the reason my patient ripped out the toilets and sinks, Bill said sardonically, "Well, that may be true, and that may not be true, but in any case it's useless."

Irv fit in perfectly with the eclectic model. He had been psychoanalyzed by an idiosyncratic genius himself, Rolf Arvidson, and had become a psychoanalyst, but Irv was in no way wedded to the old rigid ways. Like Havens, Hobson, Benaron, and most of the others, he was more a humanist, a student of human nature, than a practitioner who fit into any category or mold.

Back to Mr. Sloan. Irv asked, "So you think he might have ADD? Why?" I could tell this was not a Socratic teaching question to which he had an answer already in mind, but a genuine request for information.

"Well, the hallmark of ADD—at least what I've been taught, I'm no expert, just a rookie you know, what Elsie Freeman taught us in her lecture—is that a person has trouble sustaining attention and getting organized. It becomes a struggle to get things done and done on time, and it's not due to lack of motivation or brainpower. That's Mr. Sloan, as far as I can see."

"So what would the next step be?" Irv asked. "Don't they use Ritalin or something like that?"

"Yes, they do. I could refer him to Elsie Freeman's clinic, but she only evaluates children. I'll see if she'd make an exception, since now the experts are saying adults can have ADD."

"Good. Why don't you do that? Well, our time is up," Irv said, putting down his knitting. "As always, it's been a stimulating session. You are a talented doctor."

"I bet you say that to all the residents," I said, deflecting his praise but loving it nonetheless. Irv was always so validating. I knew for a fact that he did say this kind of thing to almost all the residents he supervised, but so what? I loved it anyway, and certainly did need it.

65.

here is her pain?" Dr. Benaron asked about Norah Devlin, the patient under discussion. Dr. Benaron and I were sitting for my supervisory session in the small office they gave her off the lobby of MMHC, a typically spare and barren MMHC office that Dr. Benaron instantly filled with her charismatic presence, as well as cigarette smoke.

Doris Menzer Benaron, one of the senior supervisors at MMHC, a grande dame of a woman, a Swiss psychoanalyst, had been close to Semrad. She was one of my favorite teachers because she was a little crazy herself, and she prized intuition, my favorite tool.

"I'm not sure," I said of the patient. "She's angry with her mother. But we are just getting to know each other."

"She has an Irish name. Is she Catholic?" Dr. Benaron asked. "Always ask about religion. It matters a lot to many of our patients and doctors usually ignore it."

"What religion are you?" I asked.

She took a drag on the cigarette that she was always smoking and then replied, "I'm Jewish, but I don't believe in God." She spoke with a German accent and delivered clipped phrases, not sentences, often separated by drags on a cigarette, followed by an exhale. "After the Holocaust, how can anyone believe in God?" Inhale, exhale. "Do you believe in God?"

"I think so," I said. "I'm not sure."

"That's good. Watch out for the people who are sure. But never forget to ask your patients about religion and God. Freud hated religion so much that he drummed it out of psychiatry. That's a big mistake because we have to look at whatever our patients are interested in. What *we* believe is irrelevant. We have to go to where they are. This patient of yours can probably talk about God much more easily than about her mother. So maybe begin by talking about God. See what she brings up. Then go there."

"She just wants to talk about how angry she is at her mother for calling the police and having her brought to the hospital."

"What else did she expect her mother to do?" she countered, sticking up for Norah Devlin's mother like the true mother Dr. Benaron was. "You told me her daughter was breaking dishes and tearing her clothes off. Was her mother just supposed to watch?"

"I know," I said. "She's crazy. Well, not so much since we gave her the meds, but she sure was crazy when the police brought her in."

"What do you think made her crazy?"

"I don't know, but the meds made her better."

"So you think it's all about the meds?"

"I don't want to think that. But it makes this job a lot simpler if you do think that way."

"You want simple?" Dr. Benaron asked. "I don't think you want simple. Do you want simple?"

"Actually, yes. But I know I can't have simple and have the truth both together. I'm just looking at where this field is heading, that's all."

"Head, shmead, I'll be dead. Meanwhile, we are where we are," she went on, lighting another smoke. "My bet is that this is a Romeo and Juliet story, only she won't let herself know it."

"Why do you say that?"

"Just a hunch, from what you've told me about her. You get to be an old lady like me, there isn't much you haven't seen. Or lived."

When she said "or lived," I was quickly reminded that Dr. Benaron herself was bipolar and had had quite a few psychotic episodes. She knew from psychosis, as she would say.

"OK," I said. "So should I ask her about Romeo?"

"No!" she snapped. "You must wait! You are so impatient. Has anyone told you that before? You are *very* impatient." She tapped her finger on the desk as she said it.

"Yes, I know that."

"Well, control yourself. It's not good to be impatient doing this work. Wait. Norah Devlin will lead you where you need to go. She may talk crazy talk but try to follow it anyway. Do you know why I say that?"

"You're trying to help me get inside the mind of the patient?"

"Correct!" Dr. Benaron crowed. "In psychosis, people think in primary process. Which can be crazy and crude. She may say fuck you, or even I want to fuck you, but don't get scared. Civilization dresses us up. To understand our sickest patients, we have to be able to think in their language. It's actually our language, too, but we don't like to admit that. We like to think we're above all that."

"I can do that," I said. "Primary process. It comes naturally to me. I like it."

"Then you will do well with psychotic patients," Dr. Benaron said, a pleased look on her bespectacled face, glasses, as usual, down her nose. "Do you know, I've been psychotic myself?"

"I'd heard that."

"Were you afraid to tell me you knew it? Does it frighten you?"

"Well, it felt awkward, I have to admit. But it makes me trust you more. You know what it's like. My dad was psychotic sometimes."

When we first met, I had given Dr. Benaron a thumbnail of my history, but that was some time ago. I should have known she wouldn't forget, at least not personal details.

"He never got over your mother," she surprised me by saying. "He couldn't accept losing her. He couldn't accept her being with your stepfather. It drove him crazy."

I was not going to cry, that was not what this felt like. It felt like someone was finally showing me what I'd been blinded to. It seemed so obvious. Why had I not seen it?

"If you're going to understand crazy thinking, this is how to do it. Try to make sense of it. It's not just all nonsense the way most people think it is. One last question. When your father threatened to hang himself that time you visited him in the hospital, why do you think he didn't? Was he afraid of death?"

"God, no," I said. "Dad wasn't afraid of anything, except maybe insulin shock. He didn't kill himself because he didn't want to do that to us."

"In other words," she said, "it was love that saved him." She paused to let me take in her words. She breathed in and out through her mouth a few times, as if she were airing herself out before changing the subject. "You were late today. Why?"

"Busy," I said. "Too busy."

"Nonsense!" she said. "Don't be so busy you don't make time to have your feelings. You could spend your whole life running away from your feelings, like most people do. I am going to train you to do otherwise. Be on time next week." She exhaled hard, as if blowing me away, and buried her face in a book she was reading while I left the bare room she filled so well.

66.

itting beside Norah Devlin I remembered that Dr. Benaron had suggested I talk with her about religion, but I was curious to hear what she would bring up first, so I first just asked her how she was doing.

Norah was a twenty-two-year-old woman of Irish parents who'd been admitted to MMHC many times since her first admission at age sixteen. The diagnoses listed in her chart included borderline personality disorder, bipolar disorder, and histrionic personality disorder. She had ripped off her clothes at home and threatened to run out into the street. Her mother had called the police, who brought her to the hospital.

In our meeting the day before, all Norah had done was rail against her mother. In this meeting she appeared more composed. She'd put on makeup. (As an intern, I'd learned about "the lipstick sign," the first sign a female patient was getting better.) She'd also done up her long blond hair and was wearing a short print dress.

When I asked her how she was feeling, she lit a cigarette and said, "Yer new at this, aren'tcha? I like that. You won't be as predictable after you've done this fer a while."

"We start out predictable?"

"God, yes," Norah said, rolling her eyes. "Soooo predictable. I've taught tons of ya."

"Maybe that's good, though—if we have a method, you know, like a predictable system for figuring out what's going on, that's good?"

"Now you sound like a priest," Norah replied, absently undoing and redoing her ponytail. "That's how they justify how predictable the services are. It's liturgy. It's tradition. It's *boring* is what it is. They bore us into submission. Can ya believe those nuns sittin' together sayin' the rosary hour after hour after hour? Don't they get a coffee break in there? You a Catholic?" Norah asked.

"No, but I know something about it. My favorite teacher in college was a devout Catholic."

"Was he now? Didja go to a Catholic college?"

"No."

"So, where'd ya go then?"

"Harvard."

"Oh, well, aren't *you* just fine and dandy. Did'ja learn anything there?"

"That teacher I mentioned. He was Irish. Well, his father was Irish. He grew up in Brooklyn. His dad was a bricklayer. He taught me a lot."

Norah let out a loud laugh. "A bricklayer taught you a lot? That must have been a hoot!"

"No, no," I said, flustered, "*he* taught me a lot."

"Which he," Norah asked, "the bricklayer or the bricklayer's son?"

"The bricklayer's son." I could feel that I was blushing.

"Don'tcha know," Norah replied, "that yer not supposed to be tellin' me all this stuff about yerself? It's quite out of bounds, ya know. Haven't they taught'cha that yet?"

And to think Dr. Benaron had advised me to bring up religion, not romance, believing it was really romance she had on her mind. As far as I could tell, she had nothing at all on her mind other than trying to make me look foolish. For people with borderline personality disorder, this was a practiced art.

"Well, my instructors are *trying* to teach me," I said, and laughed.

"Good!" Norah slapped her knee. "As long as ya can laugh, you'll do well in this game. The ones who can't laugh, they're the ones people kill themselves on."

I knew exactly what she meant. If you try too hard to keep someone alive, if you become too earnest (which I was always at risk of doing), you risk losing the person because you cease to connect with their reality. If a person says to you "Life sucks!" and you reply, "Yeah, life sucks," chances are that person will feel much better than if you replied with some reassuring statement, because in the former instance the person feels understood.

"Thank you, Norah," I said.

"For what?"

"For letting me off the hook. For telling me it's good to laugh."

"Jesus Christ, Doc, you've gotta get a grip here. I'm your patient, you're my doctor. Can we please get back into our proper roles before you drive me crazy?"

"I'd like that myself. Thanks for the reminder."

"My pleasure. Now let's get down to business. When the fuck do I get out of here? Are ya gonna try and commit me? Ya better not, ya know, or I'll make so many suicide attempts it'll make yer head spin like a fuckin' top."

"Well, you know the drill probably better than I do. We have to keep you if we think you're going to kill yourself. But it's a guessing game, and you sometimes try to mislead us."

"Asshole. Why the fuck would I do that?"

"You tell me. It's what you do."

She smiled coyly. "I suppose I do, at that."

"How can you change so quickly, from cussing me out to smiling demurely?"

"Oh, my, what a fine Harvard vocabulary. You've impressed this lovely little Irish lass."

"Now what would you call me if I called you a lass? I'd hate to think. Since when do you call yourself a lass?"

"I'm just playin' with ya, ya know that."

"My only hope is to be as dumb as I can be." The idea is to be as "dumb" as possible so that the patient can supply the facts. Dr. Gutheil had taught us this.

"Well, *that* shouldn't be too hard fer ya now, should it?"

Dr. Benaron wanted me to talk to Norah about religion, so I tried that. "You mentioned the priests and the nuns. Do you still care about that stuff?"

"Oh, my, now, you want to get deep. It's no joke bein' raised Catholic, 'specially Irish Catholic. Talk about the fear o' God, you have *no idea*. You ask me why I play all my games and why I manipulate all of you so much? *That's* why! To take my mind off the fear o' God, off o' the guilt, the shame, the burning in hell forever. Do you know what it's like to be taught from as early as you're able to understand the words that yer gonna burn in hell for all of eternity just fer bein' human and doin' what every human bein' does every day? That's what those bitches and bastards did to us. They toyed with us just like they'd been toyed with. They played with our unsuspecting little minds, and they fucked us up good 'n proper." Norah was red and shaking now. "It's a fine tradition, isn't it? Thousands of years of totally fuckin' up children in the name o' God and Jesus Christ. Gotta love it, Doc. Gotta love it."

Whew. Dr. Benaron was right. Norah did want to talk about religion. "How does it feel to talk about it now?"

"Jesus, Doc, you jus' *gotta* do better than that. 'How do you feel about it?' is *so* lame. You're in the big leagues now. Step up to the plate."

"You're a baseball fan?" I asked eagerly.

"You can't fuckin' grow up in this city and not be. Even the fuckin' cardinal goes to Red Sox games." As worked up as she had been, she was calming down quickly, undoing and doing her pony-tail again, teasing me, taking control. For such a young woman, she'd lived a long life, it seemed to me.

"Yeah. I grew up rooting for them. Ted Williams was at the end of his career when I became a fan."

"Jesus, you're *ooold*, Doc. My guy was Yaz." She turned in her chair. "I just loved Yaz. Even though he was dumb as dirt."

"Can I ask you seriously, how much does the religion stuff mess you up? Because we can work on that, you know. It doesn't have to be that painful forever."

"How do you know?" Norah demanded.

"I just know."

"Don't they teach you not to lie?"

"I'm not lying."

"But you can't know. Maybe you actually believe this, but you can't *know*."

"OK, you're right. I can't know. But I do believe it with all my heart and soul."

"OK, mister, tell me, how many fucked-up Catholics like me have you cured of all the shit we went through?"

"I'm just getting started, but others have done it."

Norah shook her head "Don't they teach you *not* to give false hope? This shit is incurable."

"Norah, give it a chance. You can change. I know you can."

She let out a long whistle. "*Tooo* earnest, Doc. You gotta tone it down. Makes me nervous."

Once again, she was schooling me. I felt bad that she had to make do with such a rookie.

"OK, you're right, I take it back. I guess what I was saying is I believe pretty much any painful memory or experience can have the poison taken out of it."

"How, with a pill?" she asked. "I doubt it. With a kiss, maybe. Maybe with a kiss."

I laughed. "You'd have to kiss a lot of frogs before you found *that* prince," I said.

"Well, you just go out and find me that one, Doc, OK?"

67.

It was a different Norah at our appointment four days later. "I'm not meant for this world," she said upon my arrival, with her head in her arms on the table.

"What happened, Norah?"

"What do you think?" she asked in a resigned way.

"I don't know."

"A boy happened, what else?" she said, her head still hidden.

"What do you mean?"

"It's still the same old story, a tale of love and glory."

"*Casablanca*. Great movie."

"But a lousy life," Norah said. "I'm so tired of it. And I'm only twenty-two. Don't think I'll see twenty-three."

"What happened? You weren't like this the last time we talked."

She kept her head down. "Yeah, that was a fun conversation, Doc. You really got me going, din'cha? Good job. Took my mind off my issues."

"I thought those *were* your issues, what we were talking about."

"I wish all I had to worry about was my Catholic guilt. Millions of people live with that without feeling like I do, you know? This just sucks. I'm so done with it."

"What else do you have to worry about?"

Long pause. "You."

"What do you mean, me?"

"Don't you see? My whole life is worrying about other people. I could be happy if I only had to worry about myself." She let out a long sigh.

I sat in silence until, to my surprise, she started to sing, barely audibly, head still on the table, the opening lines of "Bewitched, Bothered, and Bewildered."

Then she sat up. "You know that song, Doc? You will the minute I sing the next part." She continued with more of the song. She had a beautiful soprano voice. I wanted to tell her, but didn't, that that song took me straight back to my childhood because Duckie loved it and Uncle Jimmy hated it. Duckie would play Ella Fitzgerald singing it so often that Uncle Jimmy hid the record in the well of the grandfather clock in the living room and almost fired Mrs. Forgeron when she found it while dusting the clock and put it back with the other records. Instead I said, "You have a beautiful voice."

"Thanks. You mind if I finish the song?"

"Please do."

Finally standing up, she took center stage and sang the rest in a full and beautiful voice.

The song completed, she held her hands up to her face and started to cry. "Do you want to know why I know that song so well? It's because I starred in a production of *Pal Joey* and of course I *had* to fall in love with one of the stagehands. He was strong and handsome, a truly beautiful boy, and *of course* I loved him right away. I should have known he'd turn—don't they all?—but I didn't *care*. All I care is to have a man make me feel that way and make him feel that way even though one of these days it's gonna kill me."

"I don't understand, Norah. You haven't left the hospital, have you? Did you see someone?"

"Did you like my singing, Doc? Do you think I have a pretty voice?"

"Yes, I told you that."

"Men pay me compliments all the time, but that's 'cuz they want what all men want."

"You haven't told me what happened. Why were you so sad when I came in?"

"No, I haven't, have I?" She took a strand of her blond hair and drew it across her lips, then let it go. "Why is anyone sad?" she asked.

"There are lots of reasons," I said, as usual, flying by the seat of my pants. One part of my brain was trying to imagine what Dr. Benaron or Dr. Gutheil would advise me to say, while another part of my brain was trying to be real and in the moment, to empathize and understand.

"I'm tired of living," said Norah.

"And scared of dying?" I asked, taking the chance she was referring to the song.

"You're sweet, you know that, my nice new doctor? You're right, I'm tired of livin' and scared of dyin', just like Joe in 'Ol' Man River.'"

We were now both sitting at the table. Confused, I did what I was told to do when I don't know what I'm doing with a patient: I kept my big mouth shut.

"They call you Ned, don't they? That's cute. Are you gonna cure me, Ned?"

"I'm gonna try to help us both understand what's going on," I said. "Like right now, I'm curious as to what happened since the last time we met that made you so sad."

"I'm not really sad," Norah said. "Sad would actually be better. I'm just in a why-bother place. Really, why bother? Every time I do, shit happens. Can you cure me of my love of men?"

68.

ell me what happened." Dr. Benaron tapped her finger on the
desk. I sat next to the desk, she sat in front of it: we were catty-
corner from each other. "What did she say, exactly?"

"I can't remember. I asked her what she was worried about, but
I can't remember what she said."

"In the future, after the session write down as much as you can.
They're called process notes, and they can be helpful when you want
to see what's going on underneath."

"Then, the funniest thing, she started to sing."

"What?" Dr. Benaron said, surprised. "How did that happen?"

"She just started to sing, softly at first, but then she stood up
and sang the whole song. It was 'Bewitched, Bothered, and Bewildered' from *Pal Joey*. She knew the song because she'd been in the
play and had fallen in love with a stagehand."

"She sang you the whole song?" Now halfway through a cigarette, Dr. Benaron was getting excited.

"Well, she sang the song, I don't know that she sang *me* the song."

"What are you afraid of?"

"What do you mean? I'm just confused. I don't know what's
going on with her."

"You Americans," Dr. Benaron said with a bit of a sneer, "you
are so prudish and conventional. Well, you are in wrong field if you
want to be prudish and conventional."

"I don't understand."

"You don't *want* to understand. A broken leg is so much easier
to deal with."

"I'm sorry, I really am confused."

Dr. Benaron exhaled, then set to work. "If an attractive young woman stands up in a room with just the two of you there and sings you a love song, just exactly what do you think that means?"

"You don't mean— No, she wasn't singing the song to *me*."

"Then who was she singing it to, the king of Sweden? Of course she was singing to you! She was serenading you! She might as well have been doing the dance of the seven veils."

"But that makes no sense. She doesn't know me."

"Don't you understand, Dr. Hallowell, the unconscious is blind? The unconscious is stupid. The unconscious could care less about good manners. Do you want to help this woman or not? Then you have to loosen up and get into it with her."

"She is attractive."

"*Good!* I'm glad you can admit that. We are making progress now. Just because you are attracted to each other doesn't mean you have to run off into the bushes together. But it does give you an opportunity. Can you see what that is?"

"I guess to talk about what happens when these feelings come up?" I asked, feeling truly stupid.

"Yes, of course. Whatever comes up between the two of you probably comes up in the outside world, too. It's what we call grist for the mill." She paused for a moment. "Try to remember what she said when you asked her what she was worried about."

"Now I do remember what she said," I admitted. "I blocked it out because it took me aback when she said it. At the start of our meeting I asked her what she was worried about, and she said, 'You,' meaning me."

"You see how this game works with the unconscious? You're right, she doesn't know you, but we have to take advantage of the fact that her unconscious tricks her into thinking she does. She does it with every man she meets. She is turning you into whoever she needs for you to be. She is turning you into someone who can give her what she didn't get. It has nothing to do with reality. You are not going to go on a date. All you are going to do is talk. So let it happen. Remember, it is not *you* she's talking about but the person who she's turned you into, and it is not *you* who is going to help her but the process, if you allow it to."

Why couldn't other people make it so clear? Maybe because they didn't know what it's like to be crazy.

"But where are you blocked?" she asked. "Why did you not want to hear this?"

"Well, because almost the first day we got here people were telling us about boundaries and the one unpardonable sin is to get sexually involved with a patient. So I guess I was afraid of that."

"I think it's more than that. Am I wrong?"

I started to run my fingers through my hair, a habit I have when I get anxious. "It must be more than that, because I know I won't get sexually involved with her, and yet the whole scene makes me nervous."

"What's 'it'?" Dr. Benaron asked.

"The idea that she was serenading me, like you suggested. It makes me really anxious."

"Of course, part of it is just what you said, you have to be careful with boundaries, and she's a pretty woman and you're a handsome man. Do you know what Ives Hendrick advised male residents to do with female patients like this? Masturbate before you see them." She laughed. "It might have worked, I don't know."

"Who's Ives Hendrick?"

"He was a supervisor here for many years. He was jealous of Semrad because everyone loved Semrad and Hendrick was an angry, paranoid man. He once accosted Elvin in the hallway and bellowed at him, 'I'm a bull and you're a cow!' Made a total fool of himself. But he was a smart man and an excellent teacher in spite of himself. That's true for many of us," she said with a self-effacing chuckle.

I had the feeling she was giving me a rest.

But not for long. "So what is it about this woman serenading you that makes you so anxious?"

"I don't know."

"Where do you feel it in your body?"

"In my chest. Like being short of breath."

"And yet you're in no danger here. What comes to your mind as we talk about it?"

"I'm embarrassed to say."

"Well, we can leave it at that," Dr. Benaron said. "It's not good to go too fast."

"No," I said, "this is good. What comes to my mind is my mother, because that song is exactly the kind of song she loved."

"Yes," Dr. Benaron said, as if she'd known this all along. "Your mother was seductive with you?"

"Yes, she was. She didn't mean to be, it just came naturally to her."

"So now, with this attractive patient you don't know what to do with the feelings that come up."

"I guess not."

"But now they're not forbidden. They're as natural as rain."

"So, without meaning to, I unconsciously confuse the patient with my mother, get all anxious, and want to deny what's happening?"

"Of course. It's the most understandable thing in the world. We all do it. Part of your training is to get used to it."

"Will knowing this help me deal with it better?"

"What do you think?"

"I think so," I said. "But it makes me feel weird. It just seems like I'm such a cliché."

"We are all clichés, if by that you mean simple in many basic ways. And you feel weird only because you've never talked about it before. That's why we meet. To make what feels weird feel familiar. To remove these roadblocks. You're a good-looking man, female patients are going to fall in love with you, or their version of you, for a little while, and you don't want to have to get nervous and dissociate every time."

"Dissociate?" I said. "That means—"

"You pretend you're not there. Mentally, you leave the scene. It's what your mind does when it needs to escape but your body has to remain where it is."

"I guess that's what I did." I looked at the floor, needing reassurance. "Is it OK that I told you that about my mother?"

"Of course. It would not be OK if you hadn't. And remember, I'm just a crazy old lady."

"You're anything but crazy, Dr. Benaron."

"Talk more with Dr. Khantzian about all this, OK?"

I was taken aback. "How do you know I'm seeing him?"

"Because you told me!" Dr. Benaron said. "You don't remember?"

"No, I don't."

"Well, very interesting. Talk to him about *that*, too!"

69.

At one of our unit's biweekly Community Meetings, Adam, a lean and gaunt young patient, spoke up the moment the get-together was called to order by Professor Stein. Adam abruptly stood up and launched into a passage he knew by heart from the Bible:

> Hear this, you that trample on the needy,
> and bring to ruin the poor of the land,
> saying, "When will the new moon be over
> so that we may sell grain;
> and the sabbath,
> so that we may offer wheat for sale?
> We will make the ephah small and the shekel great,
> and practice deceit with false balances,
> buying the poor for silver
> and the needy for a pair of sandals,
> and selling the sweepings of the wheat."
> The Lord has sworn by the pride of Jacob:
> Surely I will never forget any of their deeds.

Adam delivered the passage with passion. He spoke the last line with a raised fist and a tremulous voice. When he finished he sat down, visibly worn out, while many of the patients gave him a round of applause.

We were all seated in the day room, where Community Meeting was always held. Before the meeting, the staff arranged the chairs in as wide an oval as possible, so everyone could see one another.

Marty, a middle-aged man of enormous girth, a very appealing wiseass, a serious atheist, and chronically psychotic, piped up after Adam spoke, saying, "I didn't understand a word he said. Did anybody understand that?"

Francie, who had been in the outpatient group Jennifer and I ran but had to be admitted to the inpatient unit because she became suicidal, said, "Yes, I understood him. Adam was saying that even though people take advantage of us, we won't be forgotten by God."

"Bullshit," Freddy said. "Bullshit" was all Freddy ever said, that and a few other stock phrases. He'd had a lobotomy decades ago, so he was more or less a broken record now. Still, I liked to imagine he timed his expostulations so that they made some kind of sense. Dr. Benaron was instilling in me the desire to decode crazy talk.

"Maybe it's not bullshit," I volunteered. We residents were encouraged to participate in Community Meeting to keep it moving along and to learn about the dynamics of a ward full of seriously mentally ill individuals. The idea behind it was that the community had its own underlying themes, a group unconscious, in a way.

"I don't think it's bullshit," said Fritz, a man of German descent with bipolar disorder. He was deeply romantic, forever in unrequited love. "We're forgotten by the rest of the world but there are people who love us."

"Like who?" Now on meds, Kenny Luongo could participate meaningfully in Community Meeting and in life. Of course, the meds were not happy pills. "The world doesn't give a shit about us."

"That's depressing," Fritz said. "Can we change the subject?"

"Life *is* depressing, or are you just now catching on to that? Life totally sucks," Marty said. "Just face it. Get used to it. Grow up, will you all?"

"You can be mean, Marty," Francie said.

"I'm not mean, I'm just a fucking realist. I don't get my hopes up so I can't ever be hurt all that much. I expect nothing."

"That, sir, is a grave mistake," Rodney intoned. In his gold vest and tan sports jacket he graced the day room with sartorial splendor rarely seen in that setting. "My paper on division by zero was just rejected again. But I will get my hopes up once again and send the paper in to another journal. No doubt it will be rejected again and my hopes will be dashed once again. But my life would amount to nothing without hope. Hope is the key to happiness."

"You are one crazy fuck," Marty said.

"Aren't we all in here?" Rodney said. "I just prefer my kind of crazy to yours."

"I didn't know we got to pick and choose," Francie said. "I'd love to have your kind of crazy, Rodney. Can you teach it to me?"

At these gatherings, John Ratey invariably stood next to the fireplace. Now he said, "There's a lot of sadness and anger in here today. Do you think it might be related to Linda's being on vacation?" Head Nurse Linda was, of course, the closest thing to a mother the ward had.

"Fuck Linda," said Stella, one of the younger patients. She had borderline personality disorder as well as a history of anorexia. Some of the characteristics of borderline personality disorder are moods that switch easily and quickly, lots of anger, insecure relationships, and intense, almost random attachments to people everywhere. Stella was very attached to Linda, so much so that she had made a suicide attempt the first day of Linda's vacation. "Fuck Linda," she repeated. "No, I really mean it, *fuck* Linda." At that, she stood up and started pacing. She lit a cigarette. "I am so fucking tired of all this, I am just fucking tired of being fucked over every day."

John Ratey intervened. "Stella, I am going to have to ask you to sit down, please."

"Fuck you, asshole," Stella said. "You can just go fuck yourself a hundred times with a big fucking dildo, OK?"

Two attendants went over to Stella. As they approached she said, "OK, OK, back off, I will sit down. OK. Just leave me the fuck alone." The attendants returned to their seats.

"Ladies and gentlemen," Professor Stein said, "what else is on people's minds? Dr. Ratey suggests we may be angry and sad that Linda is away, and we are always grateful to hear from Dr. Ratey. Who else might want to speak up?"

In the brief silence that followed, I sensed people were relieved that Stella's outburst hadn't escalated into a full-blown assault.

Professor Stein jumped back in. "In that case, I have an announcement to make. Dr. Hallowell and I have decided to start a bridge group. All are welcome. We will meet here in the day room on an impromptu basis until we find the best time for all who want to play."

"I love bridge," Freddy said.

"Freddy, you can't play bridge," Professor Stein said.

"Bullshit," Freddy said.

I hadn't expected Professor Stein to make that announcement. Although he and I had discussed it, I had not yet gotten permission from anyone to start the group. But I couldn't imagine anyone would object.

"I like bridge," Adam said. "If I am still here, I will join."

"Very good." Professor Stein stroked his beard, his version of purring.

"What do you mean, if you're still here?" Francie asked. Most of the patients were very tuned in to comings and goings and easily upset by others' departures.

"I mean I don't know how long I'll be here," Adam said. "I was put on this earth for a purpose and I need to find out what it is. Then I can leave."

"Don't hold your breath," Marty said. "Nobody here is gonna tell you why you were put on this earth because no one fuckin' knows."

"Wouldn't it be nice if someone did?" I suggested.

"Thank you for that comment," Professor Stein said. "We always like it when you speak up, Dr. Hallowell."

"What's with you and Dr. Hallowell?" Marty asked. "First you're playing bridge together and now you're patting him on the back just for speaking one sentence. You two in love?"

Lots of people laughed, a rare event in Community Meeting. I laughed as well. Professor Stein stroked his beard.

70.

Tom Gutheil taught me that my role as therapist was that of a "hired co-investigator." As I mentioned earlier, he advocated an attitude of what he called "clinical dumbness," whereby I should inquire, rather than jump to conclusions, and be as dumb as I could regarding the details of a patient's life.

Another one of Tom Gutheil's maxims was "*A* before *T*, and remember the fee." *A* stood for administration, and *T* stood for therapy. And the fee established a boundary, that this was a professional service the patient was paying for. Administration was key, especially for a patient whose diagnosis was borderline personality disorder, in that the patient had to take responsibility to earn privileges before she could engage in therapy. Earning privileges was part of administration.

We couldn't do therapy if a patient was in the midst of cutting her wrists. She'd have to stabilize first, through what are called administrative moves, like restricting her to the ward or even to a seclusion room, until she regained enough composure to be able to make use of a therapeutic conversation.

The learning curve was steep for me as a resident. Now that I had patients—real people put into my hands for help—I had to develop a systematic approach.

That's why Tom Gutheil's concise aphorisms were so helpful. They provided a guide, where I would otherwise have felt lost. My favorite of all of his many aphorisms is one I still quote often and use in my own life today: "Never worry alone."

Yogi Berra said it's impossible to hit and think at the same time. While not impossible, it's not easy to do psychotherapy and think at the same time. If you try, you become wooden and shrinklike. Patients pick up on that in an instant.

I learned to be real, but not too real. Patients didn't need to hear how my day was going, but if they asked, I tended to answer. I was learning to respect boundaries but also get close enough to form what's called a therapeutic alliance.

But my awareness of others went deeper than mere politeness and set me apart. This is something I learned from all my patients and was rooted in how I grew up and with whom. What came naturally to me, what was obvious to me, was not at all natural or obvious to others. Others needed a script. If I used a script, it came off as stilted.

What I had to train was my tendency to connect too quickly, to get too close too fast. For example, one of my first paranoid patients, Joe, and I had a good session one day. He opened up about his fears and the anger that lay behind them, about the horrible mistreatment he'd received from his father, and how much he hated the food at the hospital because it was poisoned.

"Is there any connection between your belief that the food is poisoned and what a demon your father was?"

"What do you mean?" Joe asked.

"I just mean that maybe you never feel safe, no matter where you are, because of what your father did to you."

Joe sat in silence, looking at me.

"To grow up like that, it must have left you with a feeling that there's danger everywhere, even in the hospital food."

Joe was listening, I could tell.

"So maybe if we talk about that, about how scared you felt as a child, maybe now you could realize that it's over, that you don't have to be so afraid."

"Are you calling me a coward?" Joe asked.

"Just the opposite. I am telling you that you are one of the bravest men I've ever met. I mean that." What he had endured—beatings, put-downs, daily insults, and what amounts to systematic torture—made it hard to believe he was alive today.

"Thanks," Joe said. "I think that's enough for today."

Then I made my mistake. "How about if we meet again tomorrow?"

Joe was backing up, heading for the door, when he said "Sure," and left.

That night he escaped from the hospital and took a bus to Santa Fe.

In my enthusiasm to connect, I got too close too fast. This makes anyone feel uncomfortable, but it makes a paranoid person panic. Joe panicked and bolted.

Like almost all of my talents, my intuition could be a blessing or a curse. With Joe, my having a "sixth sense," a natural ability to see into people, led me to make a big mistake and cost me and Joe what progress we might have made.

I would never see Joe again. But he lives on in my memory and I owe him for his teaching me a bit about paranoia. It would take me a long time to fully learn the lesson that it was not enough to simply declare, "Please believe me. I mean you nothing but good. If you will trust me, I can promise you I will help you find a better life." Gradually, the repeated failure of this tactic wore down my resolve to keep trying it. I grew into a more seasoned and realistic clinician. I came to accept and believe in the words of my old guide, Samuel Johnson: "The cure for the greatest part of human miseries is not radical, but palliative."

If I hadn't tried for the radical cure with Joe by naïvely believing I could gain his trust quickly, and instead opted for palliative by keeping a safe distance, one Joe could tolerate, I might have helped him instead of scaring the daylights out of him.

It remains one of the hardest lessons for me to learn, to settle for modest gains in my line of work.

71.

We residents at MMHC often worried too much that one of our patients would commit suicide. It made us defensive and overly protective of our patients. The former, illustrious head of the hospital, Jack Ewalt, the man whom Miles Shore replaced, was famous for saying that if there weren't a few suicides every year, we weren't doing our jobs right. He didn't really mean it, he was being flippant, as was his wont, but he was trying to drive home the point that if we got too concerned about preventing suicide, we'd practice overly cautious medicine and would keep patients in the hospital too long, not giving them the chance to take responsibility for their own lives.

The term "regressive," which mental hospitals are, means the environment takes care of responsibilities people normally have to take care of on their own. Semrad used to say, "Everyone wants to regress, we all do," so that the job of the hospital and its staff is to oppose that common human urge as much as possible in order to promote the skills that allow for independent living. The guessing game—and despite statistics and research, it still did boil down to a guessing game—was to decide on the earliest possible date to discharge a given patient. There's always a risk, but we tried to make it as calculated a risk as we could. Still, we residents feared we'd be blamed—implicitly or explicitly—if one of our patients actually committed suicide.

I was lucky in that none of my patients committed suicide while I was in training. My one patient who did commit suicide came to me early in my private practice. Hannah suffered from borderline personality disorder and was chronically suicidal. She was referred to me by her latest doctor who decided he had nothing more to offer her so it was time for a change, after her umpteenth admission and discharge from a local hospital.

In her midthirties when we had our first visit, she'd been in and out of hospitals since age sixteen. She was attractive still, with blond hair and some makeup, but she had the sunken eyes and pallid look of a person who's suffered emotionally long and hard. I could feel how tired of life she'd become, and yet how tough and courageous she'd been to make it this far. Fighting off the wish to die is exhausting and demands enormous strength. Where she found it I have no idea. But find it she did, year after year after year of wanting to die.

In our first meeting in my tiny rented office in Harvard Square, the then famous Fifty-one Brattle Street, she said to me, "I can't live like this any longer. I am not going to go back into the hospital ever again. I can't live my life in and out of hospitals. I don't want to and I can't." She was speaking as forcefully as she could.

"What's our plan if you become suicidal?" I asked.

"It's up to me to figure that out," she said. "I've had the best doctors in the world, I've had every medication in God's creation, I've had every form of therapy there is, and I just do not want to keep living like this. Would you?"

"No, I wouldn't," I replied. "What does your family think?"

"There's only my mother," she said. "And she agrees with me."

Hannah had made many horrific suicide attempts, mutilating herself in hideous ways that caused her enormous shame and distress. She'd been unable, despite the efforts of some great hospitals and doctors, to control herself. Today, perhaps dialectical behavioral therapy, or DBT, could have helped her, but we didn't have that back then.

I made an agreement with Hannah and her mother that I would see her as an outpatient regularly. She came to my office twice a week, always on time, and we talked about her life, her hopes, her feelings. Her mother had invited her to live with her, but she preferred shelters. Her hope was to find a job and maybe a good man.

We made a compact that if she felt suicidal, she would contact me before hurting herself. We knew that the urge to hurt herself would pass if she could wait long enough, and my job was to help her wait.

"How can we be sure you will abide by our agreement?" I asked.

"I think we both know the answer to that question," Hannah replied. "We can't be sure. I'm addicted to suicide attempts and it is up to me to break the habit."

"I hope you will try to let me help you," I said.

"So do I," Hannah replied.

One day I got a call from her mother telling me that Hannah had succeeded in taking her own life. I will never forget what the mother said to me. "Thank you, Doctor, for giving my daughter the chance to live life on her own terms. I think we all knew it would probably end this way, but thank you for giving her the chance at least to try to make a life she could enjoy."

It haunts me that today we have better treatments for what Hannah struggled with than we did then. Maybe her life could have been saved.

But this is always my lament. It's every healer's lament. If only I could have done this, that, or the other thing, the bad outcome could have been averted. Like most doctors, I have a hard time accepting that terrible stuff happens, but sometimes neither I nor anyone else can prevent it, no matter what.

Sometimes a person wants to die. Not just for a day, but week after week and month after month and year after year. One way or another, they are determined to make it happen. There's only so much we can do, or should do, to prevent it. The dicey question is how much is too much?

I knew Hannah for only six months or so, but I always found a heroic quality about her. She struggled with a terrible condition for many years, doing her best to quell the most painful and destructive feelings a person can have. She fought to live, and in the end insisted on living life on her terms, with dignity and freedom, or not at all.

It was in knowing Hannah, and others like her, people whose courage and will to live burned fiercely even in the face of massive shame and self-hatred, of abuse and deprivation the likes of which I couldn't imagine, that I started to believe again in what I'd been taught in St. Michael's Church, and taught by Fred Buechner at Exeter and in Bill Alfred's study at Harvard: We do not die, we live forever, there is life in the world to come.

I just could not imagine that the fervor with which Hannah lived, the courage, pride, and love that surged through her and kept her alive, no matter how challenged, humiliated, and defeated she felt, no matter how much the world ignored or despised her, no matter how much she despised herself, I couldn't imagine that the part of her that could hope and feel love, the part of her that could sit in my office and say I want to give life a try could ever die. It simply felt impossible, like a violation of a basic rule of physics.

72.

Like many of the decisions in my life, deciding to go into child psychiatry was not planned or well thought out. I had no idea I'd do it until the very last minute, when I had to decide on a plan for my third year of residency. The decision I did make would change my life dramatically. If I hadn't done it, I would never have learned about the condition that ended up becoming the center of my professional life: ADD.

After two years of residency in psychiatry at MMHC, we had to decide what to do for the third year. Various options were open to us. We could be a chief resident, what John Ratey had been for me when I was a first-year resident. We could do a year of psychopharmacology; a year of forensic psychiatry; a year of administrative and social psychiatry; a year of research; or a fellowship in child psychiatry, which, unlike the other options, required a two-year commitment.

None jumped out at me. But you couldn't graduate and take the exam that makes you a board-certified psychiatrist unless, after medical school, you did an internship followed by three years of residency.

Child psychiatry seemed the most interesting option, but that extra year was a turn-off, especially since I'd already been in training for seven years after college. Still, the idea of playing with children and learning about child development seemed worth an extra year. Plus child training would give me an additional calling card, an additional area of expertise, making me more employable.

Truth be told, maybe the reason for my decision was so personal even I didn't know why, as my patient Kenny Luongo so memorably taught me.

Whatever the case, that decision was to shape my career more than any other. I never could have anticipated or planned what was to happen in my child fellowship, no matter how long I thought it through or how many times I looked both ways.

73.

I didn't know what to expect when I started my training in child psychiatry. I'd had a couple of adolescents in the adult inpatient unit in my first year of residency, patients who should not have been admitted to the adult unit but had to be because there were no open child beds in the state system.

One of those kids was particularly memorable. He said to me, "People are made of three things: vision, imagination, and the lack of the first two. Most people are ninety percent lacking of the first two, which is why I don't like people."

I'd also had an interesting moment in my second year when one of my outpatients brought her ten-year-old son, Tim, with her to her appointment. Tim asked me if I could help him stop wetting his bed.

The mom had explained to Tim that I was not a pediatrician or a child psychiatrist, but he wanted to ask me anyway, since none of the doctors he'd seen had been able to cure him. He'd taken the standard medications for bedwetting and followed the various behavioral regimens, but nothing worked.

In a moment of desperation, on an impulse, I made the following suggestion. "Tim, do you know that salt holds on to water? Well, it does. What I want you to do is before you go to bed, open a bag of potato chips and eat exactly three of them. Then seal the bag and go to sleep. The salt on the potato chips should hold on to the water so you won't wet the bed." Tim nodded as I told him this, and took it in as if I were handing him the ultimate solution.

The next time I saw the mom she gave me a hug and told me Tim's bedwetting problem had been cured. The potato chips worked. Tim totally believed that the salt was holding on to the water.

It was one of my first lessons in the power of suggestion and placebo.

Beyond those two cases, I had no experience with children in psychiatry until I started my child training in 1981.

The first inpatient I was assigned in my child fellowship, which I also did at MMHC, was admitted on an emergency basis. I was paged to the unit, where I read the following brief admit note:

Mother accompanied son who was brought in by ambulance. History given by mother. 8 y.o. African American boy admitted after attempting to murder his sister by pouring lye down her throat and murder his mother by setting her mattress on fire. He himself had witnessed a murder the day before. No father present in his life. No prior admissions. No current medications or medical problems. Was medically cleared in Brigham E.R. before being sent to MMHC. Toxicology clean. Mental status delusional. Pt. is dressed appropriately but unable and unwilling to cooperate with interview. Appears to be responding to internal stimuli. Combative. Bit attendant so was put in seclusion room.

When I went to meet Tony, he looked up from the mattress on the floor where he lay, sat up, reached back, and threw a handful of feces at me. I ducked. The feces missed. I said, "Tony, I am Dr. Hallowell."

In response, the boy growled.

I said, "I'll come back to see you tomorrow," closed the door, and left.

The next day I went to the same room, knocked, and entered. Tony was sitting on the mattress this time.

"Hi," I said. "We met briefly yesterday. I'm your doctor, Dr. Hallowell."

"I know," Tony said, a big smile on his face. "I spent the night in your brain and crawled out your nose this morning."

Tony was a strikingly handsome boy, almost like a young Denzel Washington. To see him in such a disturbed state was unnerving. When kids have problems, the same problems that adults have, they're much more upsetting to everyone, caregivers included.

The Children's Unit took up two floors: the ground floor for outpatient work and the second floor for inpatients. Like the adult units, it was physically a dump. But also, like the adult units, it attracted a hugely dedicated and idealistic staff. This was always what I loved about MMHC: All of us shared a belief in an impossible, romantic cause.

Take Tony. On the face of it, he was a hopeless case, even at age eight.

Actually, Tony made the diagnostic manual—at that time, the DSM-III—look ridiculous. He qualified for so many diagnoses it made your head spin, not to mention the problems that weren't diagnoses per se but nonetheless posed huge problems: poverty, no father, hostile and dangerous neighborhood, overcrowded school, poor nutrition, minimal access to medical care, maximal access to guns and gangs.

All I could do was go where all ladders start, or in this case, where all treatment starts: the foul rag and bone shop of the heart. I had to get to know Tony.

A few days of good food, stable caretakers, a regular schedule, clean and safe living conditions, and no violence worked wonders.

Tony's apparent psychosis resolved without the use of antipsychotic medication.

When I sat down and started to talk with Tony, I didn't want to interrogate him. Playing a hunch, I asked him if he believed in Jesus.

"Oh, yes! Doesn't everybody believe in Jesus? Jesus gonna save us all. I love Jesus."

So we were off to the races. Pretty soon we'd made up a game that we called the Jesus Game. Tony and I would take turns playing Jesus, and we'd ask Jesus questions.

Tony asked me to play Jesus first. I said OK. So Tony asked me, "Jesus, what you do when someone disses your mama?"

Good question. My answer: "I'd say, 'Why do you say that? What did my mama ever do to you?'" We were sitting in my office when we played this game, usually on the floor, or sometimes we'd be outside on a bench in the park across Fenwood Road from MMHC.

"But Jesus, your mama be soooo ugly, she's so ugly she scare the pigs. What you say to that, Jesus?"

"I say, 'Why you want to dis my mama like that? She never says nothing bad about you.'"

"Jesus, I think you're nothing but chickenshit. You let me dis your mama like that and you got nothing to say but this chickenshit jive?"

Time for me to get to the point. "I don't know why you want to talk like this to me. Did somebody sometime hurt you real bad?"

Tony said, "Now it's my turn. I play Jesus!"

"OK," I agreed. "So, Jesus, what would you do if I made you really angry at me?"

"I would throw down a thousand poisonous snakes to wrap all around you and bite you and kill you, that's what I would do, and then I would shoot you a hundred times and I would throw you into the fiery furnace."

"Tony, this is Dr. Hallowell speaking. That's not the way this game is supposed to go. You're supposed to make up ways for Jesus to get angry that do not involve hurting people, let alone throwing them into the fiery furnace. Where'd you learn about the fiery furnace, anyway?"

"In church they always be talking about shit like that. How we all gonna burn if we don't be good."

"But how about if you try to play Jesus and think of ways of sticking up for your mama without killing anybody?" I asked. "Jesus told people to love each other, didn't he?"

"Yes, he did."

"OK, so now you play Jesus," I said.

"OK," Tony said.

"Jesus," I started in again, "so, I think your mama is way ugly."

"Well, I think I am just gonna have to love you for saying that because to tell you the truth she is pretty ugly." Tony laughed a long hoot of a laugh. "She's real ugly!"

"Do you love her anyway?"

"Shit, course I love my mama. Except when she whip me. Then I don't love her."

"She whip you much?"

"No, not as much as I probably need, because she's so fat she can't catch me."

"Why you think you need to be whipped?" All my years in New Orleans had trained me in the culture of African American parenting, which usually did include what they called whipping, but it wasn't nearly as severe or brutal as what white people imagine whipping to be, at least not in the hands of most mothers.

"Because I can be bad," Tony said. "I tried to hurt my sister and my mama both."

"Why did you do that?"

"Because they make me mad. I wanted them to leave me alone and they wouldn't."

"You wanna go back to playing Jesus?" I asked.

"OK," Tony said.

"Jesus, what could you have done yesterday when you were mad at your sister and mama besides trying to hurt them?"

"I could've turned the other cheek."

"Jesus, that's one of your favorite sayings, but could you think of something else you could do when you feel angry?"

"You mean besides set them on fire?" Tony said, laughing.

"Yes. Besides that. You gotta agree, Jesus, that's not very loving, setting someone on fire."

"It sure ain't," Tony said, "but it sure do feel good!"

"Jesus! That's not what you would say, I don't think."

"OK, so what does Jesus do when he gets mad? I know. He preaches. He tells stories. He overturns tables." Tony proudly folded his arms.

"Wow, you do know your Bible."

"Well, Doc, what you expect? My mama drag me to church all the time."

While I was getting to know Tony, I was also learning about ADD. In addition to the many other diagnoses Tony qualified for, ADD was on the list. The hallmark triad of symptoms that defined ADD Tony had for sure: distractibility, impulsivity, and hyperactivity.

The psych testing that had been done a few days after Tony was admitted, once his psychosis had subsided, did not mention ADD,

but that could have been because other concerns were more prominent. For example, his IQ was 69, which is borderline retarded.

I wanted to try Tony on a medication I had only read about but
never given to a patient, called methylphenidate, or Ritalin. I talked
it over with Jules Bemporad, the head of the division of child psychiatry at MMHC during my training, and he agreed it was worth
a try.

Next, I needed to get Tony's mother's permission and also run
it by the nurses and other staff on the unit. Giving any medication
to a child, especially one as young as eight, was a big deal in 1981,
and required many discussions, team meetings, chart reviews, second
opinions, parent sessions, discussions with the child, and, in general,
prolonged preparation.

Tony's mom sat in my office holding a Dixie cup on her lap that
she spat into while we talked, because she had bronchitis.

We talked about this and that for a while. I at least wanted her
to get comfortable with me as a person, not just a doctor.

"Where are your people from?" I asked, after telling her I grew
up on Cape Cod.

"Loosiana," she said. "Nawlins."

Pay dirt. "I went to medical school at Tulane."

"For real?" she asked. "My sister's babies were born at Charity."

I wanted to ask her if one of them might have been named
Fenway Park but didn't. "No kidding. That's amazing. I did my OB
rotation at Charity. That was five years ago."

"Her babies be older than that. But don't that beat all, you bein'
from Nawlins, or at least gone to school down there. You like red
beans?"

"I love them! I can even make them. Not as good as down there,
but pretty good, if I do say so myself."

"Well, Doc, I'm gonna make you some red beans and you can tell me which you likes better."

"I'd love that," I replied. "So how do you think Tony is doing?"

"How *you* think he's doing?"

"A lot better than the day he came in."

"That's good," she said, and spat into her cup.

"He's a smart kid," I said.

"How do you know that?"

"Just talking with him. He understands things quickly."

She smiled. "Yes, he does. But then why he be so bad? I watch him every day, and I try to keep him away from the gangs."

"You do a great job. And you've given him a real solid foundation in Christianity."

"I do my best. I'd be lost without the Lord so I try to fill Tony with as much of that as I can while I'm still able to."

"I do have one idea. How we might be able to help him even more. He has a condition called attention deficit disorder, and there's a medication that can help with that."

"Whatever you say, Doc," Tony's mom said. *Nobody* agrees to medication that quickly, not even now in 2017, and certainly not then, in 1981, unless they trust you. All my training in politeness really came in handy as a psychiatrist. "What's this medicine do?" she asked.

"If it works, it will help him focus his mind better and control his impulsive behavior."

"Let's hope it works," she said, with a combination of hope and desperation in her voice.

It took me weeks to persuade the nurses and other staff on the unit to give Tony a trial of Ritalin. Tony was up for it right away ("Whatever you say" were his exact words), but the staff, who were very invested in the kids on the unit and felt extremely protective of them, raised the usual objections: "Why can't we just let Tony be

Tony? Why do we have to drug him? Aren't these medications dangerous? How do we know what the medication will do?"

Once I'd addressed all of their concerns sufficiently that no one was strongly opposed, we gave Tony 10 milligrams of Ritalin, a small dose. Then we watched and waited.

The medication takes effect in about twenty minutes. Tony did not dramatically change, but staff comments later indicated he was better focused, less distracted, and less impulsive. He was also able to do better in the school on the unit. He suffered no side effects. The Ritalin didn't even cut his appetite, which it usually does.

The trial of medication seemed a success, but to get some objective feedback I asked the neuropsychologist to repeat the testing she'd done just after his admission.

Tony was on Ritalin when he had the second testing. His IQ now measured 140—a stunning increase, more than double his original score. In my thirty-six years since then, I've never seen a jump in IQ even close to that. I've told the story to neuropsychologists and some of them tell me it's impossible, I must be wrong, or the testing was flawed. But what I do know for sure is that Tony was able to use his brain far more effectively after he started his medication.

Tony also taught me how dependent your IQ is on your state of mind. When first tested, Tony was just coming out of a psychotic state, he'd been recently traumatized, he'd had little sleep, and, of course, he had undiagnosed and untreated ADD. If you can't pay attention, you will score spuriously low on all tests, including IQ.

I would work with Tony and his mom for two years, until I completed my fellowship. Tony became curious about me and my childhood. "Where'd you grow up?" he asked me.

"Mostly in a town called Chatham, on Cape Cod."

"Were you ever bad when you were a kid, I mean bad like me?" he asked.

"I don't think you're bad, Tony. You did some bad deeds, but I think in fact you are very good."

"Thanks, Doc," he said. "I think you're good, too. How come you want to work with crazy people? This hospital is for crazy people, and don't tell me it isn't, because I know it is. Why you want to work with people like us?"

"You're not crazy at all, Tony. You know that. When you came into the hospital two years ago, you were crazy, but that's long gone."

"Yes, it is," Tony said with a smile. "Thanks to you."

"And to a whole lot of other people," I said, "mostly you."

"But you didn't answer my question," Tony persisted. "Why do you like being a doctor for crazy people?"

"Well, to tell you the truth," I said, making a decision on the spot, "lots of people in my own family were crazy, so I felt a special closeness with them and wanted to learn how to help them."

"Wow!" Tony said. "That's pretty cool, Doc."

"Thanks, Tony. I think you're pretty cool yourself."

When it was time for me to leave my fellowship, Tony and I had to say goodbye. One of the rules on the Children's Unit was that clinicians are not supposed to touch the kids. Since so many of them have been sexually or physically abused, touch can be dangerous, and so it was deemed "inappropriate"—the buzzword for anything forbidden.

When we were wrapping up our last session, before I turned Tony's care over to a new child fellow, I said to Tony, "We've really come to like each other, haven't we?"

"Like each other?" he replied. "We *love* each other! I would kiss you on the mouth, but that would be inappropro."

74.

I knew I wanted to have children one day. As I've said, my most treasured dream was to create for my own children the happy and stable childhood that had eluded me. When I reached my late thirties and still had not found the right person to have kids with, I wondered if I ever would. Then a strange thing happened.

I'd always dated people for the reasons most people do: I found a person attractive, interesting, she caught my eye—the usual prompts. But this time I looked to someone else to show me the way. I didn't do it in quite as deliberate a way as that might sound, but I did do it, no doubt about that. I looked to Pam Peck to be my guide.

Pam Peck was a tall and beautiful redhead who was an occupational therapist on Service One when I was a first-year resident. Over the years I got to know her well enough to respect her a lot. Not only was she smart, she possessed a keen intuition as well as a large dose of sanity and humor, which were in rather short supply at MMHC, among both patients and staff.

She fell in love with a tall Cuban named Fernando, who was an attending psychiatrist at MMHC. Fernando fell for Pam as much as Pam for Fernando. I think it was that they were both so tall that set off the first spark. Pam just wasn't going to find many men who measured up, literally.

Attraction turned to love, which turned to commitment, and before you knew it they were married. Pam went and got a Ph.D. in psychology while Fernando advanced in the Harvard system of academic psychiatry.

Since Pam seemed to me to be one of the savviest, most stable people I knew, I looked at who she was friends with. One of her very best friends was a woman at MMHC named Sue George. I don't know how I knew this, but I did. Such is the way in institutions, relationships of all kinds become common knowledge.

I'd noticed Sue before, mainly because she walked *really* fast, scampering across the marble floor in the lobby like a sandpiper on the beach. Her nickname among the admin staff was Speedy. It occurred to me that if Pam Peck liked Sue George so much, maybe I ought to ask Sue George out. On the other hand, I knew Sue had just broken up with a man who'd been a resident a year behind me, and I didn't want to compete with history.

With these thoughts swirling around my mind, one day I walked past a car in the MMHC parking lot and saw a straw hat with a red band around it sitting on the shelf in the back beneath the rear window. The red band caught my attention. *I want to ask out whoever owns that hat*, I said to myself.

MMHC had a system by which each car that used the parking lot had to display a sticker with a number on it. If a car was blocking another car, you wrote down the number, went inside, and looked in a directory to find out whose car it was. I took down the number of this beat-up Datsun with the red-banded hat in the back and went inside to find out who owned it. Wouldn't you know, it turned out to be Sue George's car.

I went to see Laura Rood in the Quarterway Unit. As its name implies, the Quarterway was a unit for patients who weren't ready for a halfway house but didn't need to be full inpatients. The Quarterway oversaw some of MMHC's most interesting, chronic patients. Laura, the imposing, brainy, down-to-earth woman who took care of them, was a true crazy-whisperer. She "got" people with a chronic mental illness. She also got people in general, knew just about everyone at MMHC, and had a strong opinion of all of them. A lesbian in her midthirties, she was one of my dearest friends at the hospital.

"So, Laura," I asked, as she was calling to one of the patients down the hallway to pull up his trousers, "what would you think if I asked out Sue George?"

"Great idea," Laura said. "She's the best. Be careful with her, though, 'cuz she just broke up with that total snooze, but she seemed to love him, can't figure out why. Most of you men are so fucking boring, have you noticed that? Not you, of course, sweetums."

"Thanks, Laura. Coming from you, that's high praise indeed."

"Anyway, she's hurting. Just be nice."

"I'm nice to all the people I date."

"I know, I know. Don't get defensive."

Later that day, when Sue dropped by the Quarterway to talk about the family of one of the patients, Laura announced, "So, honey, Ned Hallowell wants to ask you out." According to Laura, Sue dropped her keys. "Well, that's ridiculous," she said, picking them up. "I just broke up. I'm not interested in dating."

"Well, don't say I didn't warn you," Laura said.

So runs the grapevine at MMHC. Later that same day, Laura spoke with Chris Bullock, the man who ran the inpatient unit. "Ned Hallowell wants to go out with Sue George. Why don't you have a dinner party and invite them both?"

"Excellent idea," Chris said.

Chris was good friends with Sue, as she had worked on his unit. The next day, Sue confided in Chris, "Can you believe this, Ned Hallowell wants to ask me out? I can't go out with him. I just broke up with Paul."

"You're not ready? I bet you are. Give it a try." I'll always be grateful to Chris for sticking up for me. "I think it's a great idea for you to go out with him. And it just so happens that Annie and I are having a dinner party Saturday and we want to invite Ned and you. Don't worry, Laura and a bunch of other people will be there, too."

I tried to make eyes at Sue throughout the dinner party, but she ignored me, sitting at the opposite end of the table. We didn't exchange more than two words, maybe hello and goodbye. I had done my best to talk to her before we sat down for dinner, but she kept moving away from me. Whenever I tried to get her attention, she'd look the other way.

I couldn't leave it at that. Monday morning, I walked to her office just off the main lobby and knocked on her closed door.

Sue opened the door. "Ned? What do you want?"

"Well, I'll get right to the point, I'd like to ask you out."

Sue took a seat at her desk and started shuffling papers. "I don't think so. I'm done dating doctors here."

"Oh," I said. Awkward silence. "No more doctors here."

"I just mean I'm pretty vulnerable right now."

"OK, I understand." I paused, hoping some words would come to mind, but they didn't. "Well, see you around," was all I could muster.

"No hard feelings?"

"Of course not. No hard feelings in the least . . . I hear from Pam Peck you're a real foodie. Is that true?"

"We're in a dinner group together, her and Phyllis Nobel and Sue Gorny and me."

"Sounds like fun. You meet every month?"

"Yes, and sometimes we go out for a glass of wine at lunchtime."

There was something about her. It wasn't innocence—you couldn't be innocent and work at MMHC—but she just seemed so very genuine. I said, "Well, OK, thanks for considering my invitation."

"Thanks for asking," Sue said. I bet she wanted to add, "And please do not ask again."

But I had to try once more. The next day I knocked on her door again. Again, she opened it. "Ned?"

"May I come in?"

"I guess so." Sue sat down in the chair behind her desk and I sat down as well, in the chair next to her desk, where I imagined her patients sat. I was beginning to understand what Bill Alfred meant by "the cut of her jib." I liked the cut of Sue's.

"Well, so I am here to try again. If you *really* mean no, I'll leave you alone."

Once again, Sue occupied herself with the papers on her desk. "I spoke to my dinner group about you and I was sure they'd tell me I was right, but they said I was crazy not to give it a shot."

"Good dinner group."

"But why me?" Sue asked.

"Because you're friends with Pam Peck and I think Pam Peck is about as rock solid as they come around here. That, and you have a red band around your hat in the back of your car."

"Oh, that hat," Sue said with a laugh. "How'd you know I have a hat like that?"

"It's in the back of your car."

"How'd you know what my car looks like?"

"Because I walked by this car, saw the hat with the red band, and said to myself, I have to go out with the woman who owns that hat."

"You looked my car up in the directory?"

"Yes," I said. "Very advanced detective work."

"You want to go out with me because you like my hat? I hate to tell you, but this is what I am afraid of. What happens when you discover there's more to me than a red band around a straw hat and I am not some romantic preppy type?"

"Wow," I said. "Where'd that come from?"

"It comes from dating too many men at Mass Mental and getting hurt too many times. I'm a country girl from rural Virginia, and I don't want to get hurt again by some snob who wants to be a professor at Harvard and talks circles around me."

"Seems to me like you're the one talking circles around *me*. Did you ever consider maybe I don't want to get hurt either? Or do you have a monopoly on getting hurt?"

That made her laugh. "Touché. OK, dinner. Tonight?"

"You're on!"

We went to one of the worst restaurants we could have found, ironic for two foodies, but we were rushed to find a place and a now defunct joint over the Mass Pike called Oscar's was the first place that came to mind. Burger King would have served better food.

But that night wasn't about food. We talked and talked, as people do when they are falling under each other's spell. Her mom had died of kidney disease when she was ten, so she kept her father and three siblings organized until her dad married again and another baby was born. She worked hard to get out of rural Virginia, she went to a mostly black school because her dad, even though he was a staunch Republican, didn't believe in segregation, she got in to UVA and then came to Boston for social work school. We ended up closing the restaurant and then making out in the parking lot like teenagers.

75.

My mother spent the final years of her life in a nursing home near Chatham. Between the damage done to her brain by alcohol and the falls she'd taken, she was confined to a wheelchair and was no longer herself mentally. She was still cheerful and upbeat but didn't make all that much sense.

When she died, February 9, 1988, I felt a sense of relief. I also felt guilty, as if I had somehow failed her for years. Duckie and Lyn each sent me a letter the day after Mom died.

"Dearest Ned," Duckie wrote,

After someone you have loved and has been a real part of your life for 72 years dies, it leaves a large empty space and a sense of loss that can't be filled. Her indomitable spirit will always be with us + the unhappy parts will fade away leaving us with the good memories that will never die . . . Even though, in quotes, she couldn't stand me, because I ran her life, took her children away from her + cut her down whenever we met, I felt she needed me + loved me + when the chips were down she always came to me. I was always there with open arms + tried, not very successfully, to put the pieces together again. Our relationship was hard for others to understand. It was based on love + withstood many dramatic scenes. I stayed a second fiddle to her + Uncle Jimmy for years, just to mention one + yet we kept on loving each other. If I could write a book, as thee can, the story of those brothers marrying those sisters would be more than a mini-series on TV . . .

> With love from thy ever loving aunt, Duck . . .

P.S. My letter sounds mostly about me + it's thee I'm thinking about. You tried all through your growing up years to have a loving mother-son relationship. Don't blame yourself that it didn't turn out as you hoped it would until toward the end. What a good feeling you must have when you think back on some of the lovely visits you had with her. Your love for each other healed all the hurts + it had always been there waiting for the right time to show itself. It meant a happy ending for her + rewarding feelings for thee. I love thee very much. D.

The same day, I got this letter from Lyn:

Dear Ned,

Today's my father's birthday. [Lyn's dad, my uncle Jimmy, had died nine years before.] Sort of a weird day to write to you about your mother's dying . . . What a complicated person she was. Or maybe it's the people around her who made her seem complicated. I find that I think of her in little compartments. She and I did have our good times together, mainly when we would come up to Exeter to pick you up. I think we did that twice . . . I hung out with her when I was in high school. We would have "tea" together in her various houses. Now and then, the real Dode would come out, and I did like her. Who could help it? She was funny, nice, tough, but elegant. A good person to know . . . To be related to her was maybe not so good. I realize that I carry around the reputation of "having a thing about your mother," to quote my mother. And it's true,

I sure did have a thing. But it wasn't dislike. It was really more exasperation—if you knew what she *could* be, then why the hell didn't she be that and stop being the person in one of those other compartments? Good question. I still haven't figured that out . . . I think booze had a lot to do with the failure of her to end up living happily ever after. Also, I think my parents tried to run her life for her, to take care of her, whatever, for good and selfish reasons of their own. They weren't particularly introspective. And then your father's damn illness and everyone's ignorance on that subject. Had they all lived in another time, say now, I imagine she would have had a very different life . . . How I felt about her really came from watching you try to protect her. You did a hell of a job. From the time you were a pain in the ass little kid with Uncle Unger, you protected her. During your vacations from Fessenden you were always putting up a front for all of us, who lived with her daily. You didn't fool us, but that didn't matter. You wouldn't let anyone put her down. We came to feel that it was hopeless to discuss your mother with you—and I admired that. You sure as hell cared. She was lucky to have had a kid like you, Ned. You never gave up on her . . . It breaks my heart, and I mean that, to hear you say you feel guilty. That is what I wish I had been able to take away. You really tried. You loved her so much. But you couldn't really ever reach the real Dode long enough to pull her out of those other compartments. If anyone could have, you would have been the one . . . What the hell was wrong with her, I wonder? I think basically she was incapable of really feeling anything deeply. She went through all the motions, that's for sure. But she didn't take seriously enough what most people think is most important, caring for her kids. That is the problem I have when I think of her. I just don't understand that. It obviously wasn't your fault, but I guess you'll always carry around that feeling of not being quite lovable enough . . . My father did the same thing to me to a certain extent. I have no doubt that he loved me, but it sure was on his terms, not mine. We got that very clear near his end. And that turns out to be o.k. with me. I don't want to feel much more for him. He did the best he would with what he had, emotionally. Most of it was for him. He couldn't help it, and he wasn't a bad guy . . . That's how I feel about your mother. She couldn't help it. She wasn't a bad person. She just lacked something that seemed basic . . . It makes us a kind of weird family in that we have trouble trusting that other people like us. We know intellectually that our parents loved us, at least as much as they possibly could, but emotionally we know that they basically didn't give a shit. I guess that's grounds for uneasiness! . . . So, what I'm saying is, I will miss your mother, particularly the compartment I got along with so well. I don't blame her for messing up. She played the deck she was dealt. I truly don't think there was any meanness in her at all. When she did some damage, she didn't really know what the heck she was doing. I don't think she meant to hurt you, or Ben or John, or my parents— any more than any of us really ever meant to hurt her. That people were hurt, there is no doubt. It's too bad, in a way, that we can't finger a bad guy and say he's who caused all the sadness. There just isn't a bad guy here . . . I wish she could have dropped all the compartments and just gone on with what she was—a kind, gentle, pretty lady. I wish I could go figure out what was missing—for me as well as for you. I wish I could make it so that she would never make you feel sad again for any reason at all. I wish this didn't make me cry! But I guess we are stuck with her, Ned. And there are lots of good memories, aren't there? . . . There's also this. You did a really good job of being a kid—you took care of her from the day you could walk! You really did. To feel any sort of guilt is wrong and a step backwards. Maybe feel frustrated, surely feel sad, but please don't feel guilty . . . I love you very much . . .

Lyn

76.

A year or so after Sue and I made out in the parking lot, and some three months after my mother died, I went to Shreve, Crump & Low in the Chestnut Hill Mall to buy a sapphire and diamond engagement ring. I chose that jewelry store because my mother had almost married Dick Shreve, turning him down at the last minute for my father.

A week after that, Sue and I sat at a Michelin three-star restaurant in Paris called Jacqueline Fenix in Neuilly-sur-Seine. Between the fish course and the meat course, our waiter handed Sue the following note he'd written at my request. She took it, thinking it was another wine list, and read the following message written in elegant script:

PAR CE DÎNER DU 2 MAI 1988 AU
RESTAURANT J. FENIX, À NEUILLY, J'OSE AVEC
UN PEU D'APPRÉHENSION VOUS DEMANDER EN
MARIAGE.

Sue doesn't read French, but when she got to "Mariage" she got the message, jumped up from her seat, and practically leaped over the table to hug me. "Yes! Yes! Yes!"

The other diners in this small—maybe six tables—restaurant quickly picked up what was going on and started to clap. Eating came to a brief halt.

Our waiter would have none of this, however, and quickly came to our table. "Madame, le poisson devient froid!" (Madam, the fish is getting cold!) In France, gastronomy trumps love, so Sue and I dutifully returned to our remarkable meal.

When I had told Jamie that I planned to ask Sue to marry me in Paris, it was his idea that I ask the waiter to write the proposal. Even though he was grumpy about the fish being ignored, the waiter had enthusiastically written the actual proposal.

Our wedding took place September 17 in Memorial Church in Harvard Yard, Preston Hannibal officiating because Peter Gomes wasn't available. But Preston, whom we'd not known before, was great. At the start of the ceremony he asked that the doors be kept open, since just about everyone in Cambridge had been invited.

We had eight bridesmaids, fourteen ushers, and 250 guests. I paid for it all—the ring, the trip, the wedding, the reception, and the honeymoon—with the advance I received on my first book, which was about the psychology of money. I co-wrote it with Bill Grace, a broker and the brother of Susan Galassi, who was married to my friend Jon.

Everything went according to plan, with one added adornment. Lyn dressed Ned, the youngest of her five children, the ringbearer, in a Red Sox shirt. Leave it to Lyn. As Ned proceeded down the aisle, ring neatly tied to a cushion, a cat-that-ate-the-canary grin crept across his face because he knew Sue and I would be surprised at his outfit. It was the perfect touch, added by the master of unexpected perfect touches.

At the reception, which was held at my old school, Fessenden, Lyn took me aside, gave me a kiss on my cheek, and said, "You finally did it. I love you, and I am so happy for you!"

Sue and I have been married for twenty-eight years. Lucy was born July 16, 1989. Jack came May 12, 1992, and Tucker on June 14, 1995. They were all baptized at Christ Church in Cambridge.

In addition to filling our house with children, Sue and I took various outsiders in. Maybe because Duckie and Uncle Jimmy so automatically took me in as a kid, and maybe because Sue is such a born sharer and caretaker, with the MMHC DNA in her bones, we both love to throw open our doors.

When one of our close friends was having difficulty with his young adult daughter, she stayed with us a few fun months, then went home with the mission accomplished of giving both herself and her parents a break. The son of another friend lived with us for a few months when he was looking for a job, and when Lucy graduated from college, stayed about six months. He was a football player, a three-hundred-pound jolly guy, who I worried might break the bed, but he didn't. We took in a man who goes to our church when he was having both medical and relationship problems and had to leave where he was living. He was a really smart man, interesting, raised Jewish but converted to Episcopal, and he was getting old. I think he stayed maybe a month or so. We were sad to see him leave, but happy he'd found both a new place to live and a new love.

But the longest stay with us, by far, was by a man I hadn't known at all nor had any connection with until we met. One day he came to my door looking for help with his ADD. I told him I had an office where he could see me, but he asked, since he'd come all this way, couldn't I just see him now at home. I told him I couldn't see him as a patient in my home, but I could talk to him as a fellow traveler on the road of life, or something like that.

I invited him into our living room where we sat and talked for an hour or so. His name was Rex. He was in his late thirties and had moved east from his family home in Chicago to stay with his sister near Boston. A skilled handyman, he planned to move in with his sister and her husband and help renovate their house.

Rex was a tall, good-looking guy with angular features. He looked the part of a carpenter—jeans, old work shoes, faded work shirt—and he had the natural buoyancy and joie-de-vivre that so many people with ADD exude. He'd found me because he'd read one of my books. While working for his sister, he also did some odd jobs for a family in Winthrop, and as payment they'd given him an old beat-up car, which is what he drove to my house.

Turned out his sister had given him $10,000 when he started working for them, which Rex used to pay off bills he owed. Now, seven months later, his sister had not paid him any more, even though he'd been working twelve-hour days renovating her house. The friend he worked for in Winthrop told him he thought he was getting shafted.

"I had this teacher," I said to Rex, "who used to tell us a Russian proverb. 'Be a sheep, and a wolf will appear.'"

Rex laughed. "Well, I've been being a sheep for sure. The problem is, I don't have anywhere else to live."

"So you have to let her exploit you like this? Tell you what, if you can't make a better deal with her, and if my wife agrees, which I am pretty sure she will, you can live here until you find a steady job and can afford a place on your own."

I knew Sue would agree, but I didn't expect Rex to follow up, for any number of reasons, not the least of which was that he had ADD. However, I was wrong. He moved in with us, lived in our third-floor bedroom, which has its own bath, and stayed for a year. He did odd jobs around the house in exchange for his room. He moved out when he found a job that paid him well enough to rent a place to live, but in a few months he lost that job, so he moved back in with us for another six months.

Now, nine years after I first met him, he's been employed steadily for the past seven years, is living with a woman whom he loves and who loves him back, and has saved enough money to buy a house.

Maybe Sue and I are just part shepherd. Les Havens—who "got it" when it comes to being human better than any teacher I had— told me this quality would make me a valuable human being but also put me at great risk in this world. He said not to worry, though, because there was nothing I could do to change it anyway, and the world, which would never pay me back for any good I might do, and very well might punish me for it instead, was better off having me just like I was. Havens, who believed in God but rarely spoke of his faith, said to me, "You give, they receive. But St. Francis got it right. In giving, we receive. That's why this is the best job in the world. But don't you dare tell anyone I said that or I will lose my reputation as a sophisticated, cynical Harvard academic."

77.

After my child psychiatry fellowship ended, I was offered a job as an attending physician on one of MMHC's inpatient units. The position had been called superchief when I first arrived. Bill Beuscher had been my superchief; I thought I could pass along to others what he'd given to me. Though the pay was low, I'd long since bought into the romance and mission of this jalopy of a hospital, so I gratefully accepted the job.

The man who hired me, Dr. Miles Shore, became head of the hospital in 1975. He had two titles, reflecting the dual oversight of MMHC. The Massachusetts Department of Mental Health named him superintendent for running a state hospital; Harvard Medical School gave him an academic title, Bullard Professor of Psychiatry.

As head of a psychiatric hospital, as well as a professor at Harvard, Miles was loved, hated, admired, scorned, envied, pitied, relied upon, competed with, or ignored, depending on who you were. Dr. Benaron said to me, "Miles killed Semrad." It's true he did not put Elvin Semrad on the pedestal Dr. Benaron did, but Miles was also trying to bring MMHC into the contemporary world of psychiatry, with its emphasis on the diagnostic manual, medication, community-based interventions, biology, genetics, epidemiology, social psychiatry, and a variety of other topics Dr. Benaron was less interested in. She championed the heart, and innovative ways of reaching people who were hard to reach. I actually don't think Miles disagreed with her. He was just trying to bring the hospital up to date.

Les Havens was not as blunt as Dr. Benaron—few were—but he also didn't love Miles. When he'd had enough, he left his beloved

MMHC for Cambridge Hospital. In a one-for-one trade, MMHC acquired the estimable George Vaillant from Cambridge.

Miles did not have what so many of the prior greats at MMHC had in spades: charisma. Like his name, Miles was nerdy, tidy, and—to use a word I deplore—appropriate. I found him to be exactly what MMHC needed. Just as my family had found a welcome dose of normality in Tom Bliss, MMHC needed a dose of stable, by-the-book, research-based psychiatry. Above all else, Miles was *sane*. It is better to have a sane and stable person in charge of a mental hospital than a charismatic, unpredictable star.

Dr. Shore gave me the job, but that didn't mean he loved everything I did. For example, there was the cat.

First, some background. When I took over the unit, Alan Brown was chief resident, overseeing a group of six residents and two psych interns in their first year. Alan was a gifted, brilliant man. Tall and thin, with a dark beard, Harvard undergrad and Yale medical school, Jewish parents and a stunningly beautiful WASPy blonde nurse practitioner wife and two little kids at home. Knowing suffering in the Jewish tradition, he knew humor even better.

More by-the-book than me, Alan was nevertheless willing to try new approaches. I had been one of his supervisors the previous year, and we often laughed about a married couple, Marcia and Jay, he presented to me. "They say they feel uncomfortable meeting in my office," Alan told me. "They're both pretty thought-disordered, and the confines of my office make them paranoid."

"Have you tried leaving the door open?" I asked. "That can help."

"Yup, tried that," Alan said, shaking his head.

"Have they said where they might feel more comfortable?"

"Well, yes," Alan replied, "they said they'd like to meet in the main cafeteria up the hill at the Brigham. But that's out of the question, of course."

"Why?" I asked, as usual pushing the bounds of conventional wisdom in favor of innovative treatment. "I know it sounds off the wall, but why not go where they feel most comfortable?"

Alan laughed. He pulled on his beard. He always tried to treat me with respect, but I could tell this idea was putting his forbearance to the test. "You want me to meet with these two thought-disordered patients in the Brigham cafeteria? Surrounded by the general public? You actually think that's a *good* idea?"

"I know it's unusual, but why not give it a try?" I replied. "What's the downside? Are they at risk for becoming violent?"

"Well, they never have been violent, but then again, we've never met in the Brigham cafeteria surrounded by total strangers!" he replied, slapping his knee for emphasis as he laughed.

The following week Alan and I met again in my office for supervision. He had reluctantly taken my suggestion and met with the couple in the Brigham cafeteria. It turned out *not* to have been such a great idea. His description of it still makes me laugh out loud.

"So there we are at a table, just the three of us, right in the middle of the hubbub of the Brigham cafeteria. Right off the bat, some cardiology fellow comes by, I could tell by her ID, and asks if she could sit down with us. 'Sure!' Marcia says. I practically tackle the cardiology fellow trying to stop her, saying, 'I'm really sorry, but we are having kind of a private meeting right now.' 'Oh, I'm sorry,' says the fellow, looking quite nonplussed by my near assault, and goes away shaking her head, like what is this world coming to, looking for another table. 'Why did you do that?' Marcia demands of me.

" 'Well, I thought we needed some privacy for our session, don't you agree? Do you both feel more comfortable here than in my office? Should we just move back to my office, maybe?' I ask, almost pleading.

" 'Feels way better here,' Jay responds, '*way* better,' and then in a full, loud voice announces, 'Marcia won't give me as much sex as I need! I need more sex!'

" 'You're just a *fucking* machine,' Marcia replies, in an equally loud voice. 'Sex, sex, sex, it's all you ever want!'

"Of course, I am looking around like I have ants in my pants, praying no one is hearing this, but knowing that they couldn't *not* be hearing it. I'm about to crawl under the table when Jay suddenly blurts out, 'Marcia, why don't we do it right here, right here on the floor in the cafeteria? Wouldn't that be far out?'

" 'Jay, you are such a romantic,' Marcia replies, 'it's what I love about you, but no, honey, I don't think we should do it right here, we could get arrested, and I'm pretty sure Dr. Brown wouldn't approve, am I right about that, Dr. Brown?'

"Of course, at that moment I want to hug Marcia, but I don't. I'm racking my brain trying to think of a nice, quiet topic we could discuss to calm down this conversation, so I ask, 'Are your meds at the right level for each of you?' I knew they were because I'd just checked, so I figured this was safe.

" 'Fuck the meds!' Jay said, slamming his palm down on the table. 'Fuck the fucking meds. I want to fuck on the floor.'

" 'Honey, Jay, calm down, OK, we gotta be presentable here, or Dr. Brown won't meet with us here anymore.'

" 'That's right. Inside voices,' I said reflexively, as if I were talking to one of my kids. Anyway, with Marcia's help we survived without getting thrown out by security. We talked about meds, and then I bribed them with an offer of buying each of them a coffee and a hot dog if we could take the session back to my office. I know we're not supposed to bribe patients, but desperate times, y'know."

In my office, I apologized through tears of laughter. "That was my mistake. I just thought if they wanted to meet in the cafeteria, why not?"

With a wide smile, Alan said, "I guess we found out why not. I'm glad I did it, though. Truly I am. I will never forget it, and I never would have done it without your encouragement."

"I don't know if that's a good thing or a bad thing," I said with a slightly apologetic smile.

So when Alan and I started running the unit together, we already knew each other well. We did our best to teach the residents what we thought they needed to know, and to help them get along with the staff. Alan did a great job.

We both liked to innovate. For example, Alan started a "No" group. This was for all the patients who rejected out of hand the idea of psychotherapy of any kind. It quickly became the best-attended group on the ward. Without knowing it, the patients were getting the benefit of a group—sitting with other people and communicating—while protesting the usefulness of a group, or any other form of therapy. The "No" group was a huge success.

Back to the cat. For my part, I'd always wanted to bring an animal onto the ward. My preference would have been a dog, but dogs need too much care and attention for a psych unit.

I thought an animal could connect with the many cut-off patients we had—the schizophrenics, the quietly psychotic—as well as pretty much everyone else, including staff.

But when the people in charge of risk management got wind of my idea, you'd have thought I was proposing bringing in Ebola-infected bats.

The objections began with medical issues, and escalated from there. "What if a patient is allergic? What if the cat brings in a dead mouse and the patients freak out? Do our insurance policies cover cats? What about cat-scratch fever? What if the cat dies? What if patients get jealous over who the cat loves most?" The objections went on and on . . . and on.

Finally I just went out and got a cat. Alan and I announced at Community Meeting that we now had a cat on the unit. Jake, the man in admin who'd raised the most objections, said to me, "You had no right to do it and you will regret this. Mark my words."

We had a contest among the patients and staff to name the cat. I liked Alan's suggestion best: Eddie-puss. But another name, Mister Fenwood, won, after Fenwood Road, the road the hospital was on. Unimaginative, but serviceable.

The patients *loved* Mister Fenwood so much that the cat quickly got fat from being overfed. Patients took turns emptying the kitty litter; they held bake sales to raise money to pay for the litter, cat food, and vet bills. Later, they helped supervise each other to prevent overfeeding.

One day, however, despite the patients' supervision, the cat fell out a window that had been inadvertently left open. The unit was on the third floor. Luckily, Mister Fenwood did not die, but he did break his hip. We had two options: We could either put the cat down or pay three hundred dollars to have the hip surgically repaired. The patients' bake sale fund had a balance of only about twenty dollars.

Jake, the chief admin, could hardly conceal his delight. "I told you this cat would lead to no good. Now you have a bunch of unhappy patients on your hands and you have a cat that has to be put down. What are you gonna tell them? That you got carried away with another cockamamie idea of yours and weren't smart enough to say no when you should have? I'm tellin' ya, Hallowell, this never should have happened. It's all on you."

"Don't look so happy," I said.

"Just had to say I told you so, that's all."

I ended up paying out of my own pocket to have the cat's hip repaired. The patients spent an entire community meeting discussing safety procedures for the cat once he returned from the hospital. Mister Fenwood lived to eat again.

Anyone who visited the unit had to be impressed with what a masterful therapist Mister Fenwood became. He might as well have been Board Certified for all the good he did. Patients who would not talk to anyone else would sit with him on their lap, stroking him, talking to him, connecting with him as he purred. Mister Fenwood would make daily rounds, visiting various nooks and crannies, curling up on chairs and in the odd corner, nuzzling one person after another, and, of course, taking long naps.

Jake was not a bad guy. I know I was a thorn in his side, a pain in the ass. His job was to minimize risk, and I was forever proposing ideas that only increased risk and were not at all necessary in delivering the accepted standard of care. If I were Jake, I would have stuck pins in a doll of me for sure. Because I didn't stop with just the cat.

Taking a page from the *Cuckoo's Nest*, I organized a trip to Fenway Park for a group of patients. Jake couldn't stop sputtering when I told him about it—was I "fuckin' *crazy*?"—but after the game, when I described to him how Professor Stein had carefully explained the rules of baseball to a group of Asian fans sitting next to our group, he had to agree that the idea had worked. "But you got *so* lucky. It could have been a total disaster *and you know it!*"

I countered, "So could group walk be, when they cross the street. Should we eliminate group walk?"

"Of course not. But why stick our necks out further than we need to?"

"Because we want our patients to realize they can do more than sit in the day room and smoke cigarettes," I said. "We really ought to have them working a farm instead of being cooped up in this warehouse."

"It's not a warehouse, dammit, it's a Harvard teaching hospital."

"Exactly, which is why we're trying to encourage innovation. That's what Harvard is supposed to do. Lead the way. You got behind Miles's program to innovate with the Fenwood Inn, why can't you get behind the little things I do?"

The Fenwood Inn was indeed an innovative program Miles Shore had developed with Jon Gudeman in which patients who were not critical but still needed to stay overnight could be housed but not require the staffing of a medical unit, nor did they have to abide by the myriad regulations on such units. Miles was justifiably proud of this contribution to public sector psychiatry.

Jake wasn't convinced by my argument. "Miles—that's Dr. Shore to you, son—runs this place. He is the Bullard Professor of Psychiatry and you're just a peon."

"Tell me about it," I said.

Jake and I actually liked each other; we were just advocating different interests, and we enjoyed sparring.

One day the Boston *Globe* came to visit. Dr. Shore was promoting the Fenwood Inn, and he hoped the *Globe* would give MMHC some much-needed, well-deserved publicity. The reporter and photographer visited the hospital for a half day, and, by all accounts the visit went well.

Dr. Shore, however, was not altogether pleased. He stopped me in the hall and said, "Hallowell, the *Globe* was here today, and all they wanted to talk about was your fucking cat!"

78.

One late afternoon in the dead of winter, Professor Stein veered off from group walk. He had unlimited privileges, and sometimes, rather than staying with the pack, he'd go off on his own.

The pack itself was a diverse group of six or eight chronic mental patients strolling down Brookline Avenue. They'd talk to one another, remind one another to look both ways while crossing the street, tell the leader what time it was to make sure they got back before the dining hall closed, and meander along the sidewalk, sometimes turning in circles while looking up at the sky, or stopping to pet a dog who was being walked past them, or leaning down to inspect something on the sidewalk that caught their attention, always just casually ambling, without the determined pace of the average pedestrian.

These pedestrians were teetering on the slim boundary between being out of their minds and connected with reality. Going for a walk in the supposedly sane outside world was one way we tried to help them adapt to leaving the hospital completely one day. They might be lost in their thoughts, or quite in the moment: enjoying being outside, enjoying stopping at the 7-Eleven to buy smokes, candy, magazines, or coffee, and relishing those treats far more than the average person does.

It was as a result of one of those group walks gone wrong that I was conceived, when my father was allowed to visit home while actively psychotic. Of all people, I had to appreciate the unpredictability of what can happen on a group walk.

You simply cannot supervise the mind all that closely. Sanity is elusive and fleeting. You never know what a mind is going to have it in mind to do. Try as you might, with all the knowledge, medications, rating scales, checklists, background information, and past history that you could ever want, you just never know what's gonna happen next.

It's why our diagnostic manual, such a brave attempt to classify the peculiarities of the mind, is in the end such a paltry thing. There is so very much more in heaven and earth and in the human psyche than was ever dreamed of in the diagnostic manual.

And praise God that it's so. Though many claim otherwise, the mind is, at its center, free. Free to go where it wants, dream what it likes, get lost in places unknown, entertain terrors and joys of all kinds, and puzzle us all, especially itself.

It should not have surprised me as much as it did to learn that Professor Stein had taken a long detour that frigid evening. He stopped at a package store on Brookline Avenue, bought a bottle of Jack Daniel's, and found his way into some snowy woods where he sat at the base of a tree, leaned back against the trunk, drank his bottle of Jack, passed out, and died.

Everyone wondered why he did it: Was it a suicide or an accident? I think he was tired of life. Professor Stein was an intelligent man. I think he wanted to die. He knew that passing out sitting in the snow in a deserted wood in subfreezing temperatures was likely to be lethal. As much as he enjoyed our games, how much could playing bridge do? With no friends or family, aside from the people at MMHC, as well as a mind that rarely gave him a moment's peace, how much of his existence could make him want to keep on living?

Still, the news hit me hard. I had come to love Professor Stein, with his nicotine-stained beard and his wise, stabilizing presence on the unit.

I wanted it not to be. At that point, I turned to my old friend Samuel Johnson, taking one of his books down from the shelf in my office.

I read the final stanza of "The Vanity of Human Wishes," a poem hardly anyone reads anymore except in a college course, but one that speaks to me in a unique way, loaded with powerful personal associations as well as Johnson's clear-eyed, worldly warmth.

Decades later, on an Easter Sunday, I attended services at Christ Church with Sue. That church is where Lucy, Jack, and Tucker were all baptized and confirmed, where I eagerly go to get a much-needed serving of spiritual food. Bringing all my troubles, worries, and concerns, as well as joys and celebrations, here I feel a cozy and abiding community. Sue and I go every Sunday we're in town. I go because I have to; I'd feel depleted if I didn't.

Christ Church is a collection of wildflowers: street people and Harvard profs; radical politicos and old-money WASPs; a college classmate of mine who is black, gay, and the former mayor of Cambridge; old people in wheelchairs next to babies nuzzling in their mothers' arms. The congregants enjoy a wide diversity of religious beliefs, disbeliefs, and hunches. You don't have to believe anything to be a member of Christ Church. All you need is to show up.

Abutting Harvard Square, just across Mass Ave. from Harvard Yard, the church bursts at the seams on Easter. Easter is usually a crisp and sunny April Sunday, and this was no exception, with everyone decked out in their brightest finery, including me wearing my collar pin.

Built in 1760, Christ Church, painted light gray, is simple in design, with Doric pillars and multipaned windows lining both sides of the nave. On clear days, lavish sunlight and a gentle breeze fill every pew. Two glorious, sparkling chandeliers hang from the high ceiling. The choir sits behind the altar, with the organ console in a cozy pit dividing the two.

Our church has three gifted priests, a phenomenal music director, Stuart Forster, as well as an exceptional organ, and an exceptional volunteer choir decked out in their red robes and white surplices. With the addition of a couple of trumpets, the Easter service fairly lifts off the roof, ringing jubilant music throughout Harvard Square.

On that particular Easter Sunday, when Stuart started to play the lead-in to "Jesus Christ Is Risen Today," a hymn I know as well as I know my own name, I filled with emotion as if a dam were breaking. The hymn itself is a heart-thumper, but that day it was amplified by our organ, by trumpets and choir, and by the over-flowing congregation singing their loudest. Add the sunlight pouring into the room, and that we were all there to celebrate nothing less than the biggest deal of all—resurrection, eternal life, the paradise that we're told awaits us—and you feel God right there with you, as if God were on the brink of bursting out of the invisible. For some, now and then, He actually does. Are you crazy if you see the face of God? Short of that vision, if you feel God's presence, as I did then and so often do, as persuasively as the sun on the back of your neck, does that mean you lack the courage to accept the nothingness some say awaits us all?

For reasons I can't explain, I have come to cherish and trust the feeling of God's presence. For me, it is as real and present and impor-tant as love. It's always there, sometimes like a nagging question, sometimes like an abiding belief, always dynamic and in flux, but never irrelevant or absent.

At that moment on Easter Sunday, oddly enough, I thought of Professor Stein, an avowed atheist, sitting against his tree in the snow drinking his Jack. I imagined the warm buzz he felt as the bourbon did its work, pictured him falling asleep as the bottle dropped from his hand onto the snow beside him. I wondered if he'd been able to finish the bottle. I hoped so. I wondered if he were with God now.

Please be, I prayed. *Let it be so.*

To watch us sing and celebrate in church that day, Sue and me and the hundreds of others, you would have thought we all *knew* it was so, that we harbored no doubt that Christ our Lord was indeed risen that day, as would all of us rise when our day came. At Exeter, Fred Buechner had found all manner of poetic ways of telling us that he, also, was sure it was true, and that doubt was just one part of the truth. Harvard prof Bill Alfred had told me *he* knew it was so, just as sure as he knew his Franklin stove sat in the corner of his study.

How much better it would have been for Uncle Jimmy, who was so terrified of dying, and Samuel Johnson, who was even more terrified, and the countless others of us who can't find any peace around the fact of death, to know that we do *not* die, that after death there is not blackness, that we do not lie in cold obstruction and rot, that there is life for all of us, forever.

My spiritual journey began all those many years ago saying the Lord's Prayer with my mother in bed at night. Now it had come to this: an Easter morning in church, when by rights—and by statistics—I should have been dead or at best down and out on the margins of life. Instead, I'd witnessed more than my share of comebacks in a myriad of times and places, including achieving my own most cherished goals. I had no way to explain any of it, no proofs, but the tools I trusted most, my intuition and my ability to apprehend in the dark, fallible as those tools can be, still led me to plant my flag, for better or worse, right here where we stood, on the miracle some call a fairy tale, the miracle we were all praying, singing, and swaying together like a field full of wildflowers in a breeze to celebrate on this Easter day.

Fools perhaps we were, but fools for love.

79.

Sometime in the early 2000s, Lyn went for a routine physical and it was discovered that a marker of inflammation was elevated. One test led to another, then another, until the diagnosis finally came back: multiple myeloma.

Lyn and Tom came up to Boston to see one of the world's leading specialists in multiple myeloma, Ken Anderson, professor at Harvard Medical School and the Dana Farber Cancer Institute. He laid out a plan of a series of interventions, which held great promise.

Over the course of the year, Lyn endured procedure after procedure, all the while reassuring us that she'd be fine. I knew she didn't want to discuss possibly bad outcomes. She simply said, "What will be, will be."

When a stem cell transplant didn't pan out, we began to fear the worst, even as Lyn told me over the phone, "I'm not going to die."

"Of course you're not," I said.

A few months later, it was in 2002, I was somewhere in Vermont, having given a talk in a school. Standing outside in the freezing cold, I spoke to Tim on my cell phone. He said, "Mom is not doing well."

"Oh, no, not again."

I could tell from Tim's voice this was more serious than the many other setbacks. "She's back in the hospital," he said.

"Is she . . . I don't know . . . Could she actually die?"

"Yes." Tim's tone suggested to me that he'd hoped I'd ask so he wouldn't have to volunteer that information. "It looks that way."

What happened next remains a blur. I drove straight home, talking to Sue along the way. After I picked her up, we drove to the Rhode Island Hospital in Providence.

When we arrived, Molly was there with Tom. Molly had spent the previous several weeks lying next to Lyn in bed, stroking her mom's back, rubbing her feet, comforting her any way she could.

As I stood at the foot of her bed, Lyn looked up at me. "I don't want to die. Neddy, could you please do something about this?" She had almost no energy. It was clear the time left was short.

I went to the hospital chapel and did the only thing I could do to try to fulfill Lyn's request. I said a prayer.

She died soon after.

She was fifty-eight, Jamie was fifty-six, I was fifty-two. We'd been such a great team for so long. Her dying hit me hard. For Tom, Molly, Tim, Jake, Anna, little Ned, and everyone who knew her, her dying was beyond tragic, it was impossible to comprehend.

Lyn was simply not the dying type. She was the living type, oh boy, was she ever. You could not be in her presence and not get ready—for a zinger, a kind remark, an amazing insight, or a call to action.

She could not be ignored.

Perhaps that's why none of us could believe she had died. Tom would say she was just in the next room taking a nap. Most of us would get hit at some point on certain days with the reality of her not being alive any longer after finding ourselves not believing it or forgetting it. It must have been a mistake. Check the manifest. You picked up the wrong body.

Even to this day, fifteen years later, I think it's all a dream that I will wake up from. She will be alive and I will say, "You won't *believe* the terrible dream I just had." So far, I haven't woken up from that dream.

We found our ways of getting on with life.

Tom Bliss continued to be the great orthopod he always was but gave up surgery after a few years, focusing on outpatient work. He misses the O.R. but found a new operating theater, the world of photography. He won't sell his superb work—that's not why he's doing it—but someday I think a Bliss photo will be worth a lot of dough. He found himself a new woman, Mary, the niece of Eileen, the Irish lady who'd been just about everything to the Blisses for many years, from nanny to babysitter to adviser to honorary member of the family.

Smart and kind, Mary is a professor of social policy and intervention at Oxford. She still lives in Oxford but comes to the United States regularly, as does Tom go to England. In Provincetown, the two of them bought a house together, where they go in the summer, and sometimes in the winter.

Five or six months after Lyn died, Sue and I were fooling around in bed, kissing and such, when out of nowhere Sue looked up at me and said, in a clear and unmistakable voice, something she'd never said before: "Pecker up!"

Holy fuck! I said to myself as we both abruptly stopped what we were doing. "What did you just say?" I asked, while Sue, dumbfounded, simultaneously blurted out, "What did I just say?"

"I think I said 'Pecker up,'" Sue said. "What does *that* mean? Why did I say it?"

"'Pecker up' is *exactly* what you said," I replied. When I told her the story about the promise Lyn and I had made, even Sue, ever the skeptic, had to acknowledge it was at least spooky. She couldn't go so far as to agree it was proof of an afterlife and that Lyn was firmly ensconced in paradise, but she had to agree it gave her reason to wonder. She didn't know the phrase "pecker up," she'd never used it, and she had no idea what it meant. The words simply fell out of her mouth. "Well, I must have heard it somewhere before" was her explanation. "But you are right, it is pretty strange."

I smiled down at Sue and up at Lyn. I'm sure she, like Sue, would have found reason to discredit the evidence she had just supplied me through Sue's lips, but that was fine with me. She was safely tucked away in eternity. I had the proof I wanted. Now I just had to believe what I'd seen and heard and hold on to it through my inevitable, irascible periods of doubt. I just wished I'd had a more articulate response to this communication from beyond than "Holy fuck!"

On the other hand, given the raft of improbabilities that had followed me from the moment of my conception, maybe those two words summed up my life. As Sue drifted off to sleep, I lay on my back staring at the ceiling, knowing that I, like the sailor who saw Icarus fall into the sea, had at least just been part of something amazing.

I lingered over one fact I knew for sure, one fact I would return to and savor many times for years to come: If Lyn had not been there for me, I never would have found my way. "What will be, will be," she used to say. And yet she never accepted what was, but spent her life working to improve the lives of all the people she loved. She pretty much saved mine.

There was a lot that needed saving in my crazy family, but I'm proud of the whole lot. We didn't know what we were doing much of the time, we made stupid plans if we planned at all, we stumbled into situations we didn't expect, we blew fortunes we could have made, and we often failed to deliver on the potential we were born with. But we were true to ourselves, mostly because we were simply not able to fake much of anything for very long.

Dad was a champion sailor, racing from age fourteen and winning a slew of trophies at the Wianno Yacht Club. One of his special talents was finding wind where no one else could see it. He would veer off from the others and find the gust that would win him the race. A version of that talent landed in us all. Somehow we found what wind we needed in corners of the bay where no one else looked. We found the wind, we found a course, we navigated in our own fashion, and we helped one another out the best we could.

EPILOGUE

Since this has been a storybook, I thought I could fit in one more story, a quick one.

As I said at the start, no one can tell his entire story. In this account, I've left out the writing of my books, and my work with the fascinating condition misleadingly called attention deficit hyperactivity disorder, or ADHD, which I discovered I had in 1981.

I've left out hundreds of stories of patients I've worked with, friends I've made, and crucial world events and scientific developments that shook me, shaped me, and changed our world.

As I was wrapping up the writing of this book, I was also seeing patients in an office on West Seventy-Second Street in Manhattan. I took breaks between scheduled patients, free hours during which I could put the final touches on this work.

It was April, with the weather changing from cold to chilly to warm enough to go without a jacket. On one of those sunny, warm-enough days, I decided to go out and get a dose of my medication, aka Starbucks.

Walking down Columbus Avenue, white Venti cup in hand, I saw a man sitting on the sidewalk up against a streetlight pole, lovingly stroking a big, black dog, could have been a Lab, more likely a mix. Since I love dogs, I stopped for a moment and watched. From the looks they gave each other, I could plainly see how much the man loved the dog and the dog loved the man. This is life at its best, if you ask me.

Except that the man was down and out. I couldn't tell if he had two legs, one leg, or none. He was dressed all in black, and the covers he had around him concealed his full anatomy. Looking for money, he had placed his paper cup on a larger piece of cardboard I suppose he'd torn from a box.

After enjoying the warmth of man and his dog for a few seconds, I started walking toward them again. As I got closer I saw he had written in clumsy caps on the cardboard ANYTHING HELPS.

With my coffee in one hand, I couldn't get my wallet out of my hip pocket and open it to find a bill, so I asked the man to hold the cup for me. He gladly obliged, smiling at me, along with his dog, who also smiled at me, as dogs do.

That always gets me: They're smiling, I'm hassled, what's up with me? I took from my wallet a five-dollar bill—Why not a twenty?—put it in his cup, took my coffee back, and walked on. He said after me, "That's a large amount, thanks so much."

When he said that, I almost turned around and went back to get the story of his life, find out more about his dog, maybe instead give him the twenty, and do all the other stuff I am so inclined to do in spite of the many good reasons not to. But I didn't. It didn't even occur to me at the time that he was, by all rights, where I should have been.

Maybe if I see him there the next time I'm in New York, I will talk it all over with him, find out his story, tell him mine, and make friends with his dog, of course.

ACKNOWLEDGMENTS

I want to thank all of the characters who appear in this memoir, some disguised, some without disguise, just as I remember them. These are the people who made my life what it is, so I owe them the world.

I also thank the professionals who helped me so very much, my agent, the brilliant and savvy Jim Levine; my editor, the astute and dedicated Nancy Miller, and her whole team at Bloomsbury who punch far above their weight; my personal assistant, the incomparable Dianne Nargassans, who saves my bacon regularly while also reading Epictetus daily; my media assistant, the adept Christina Veal, who negotiates social media for me; and my partner Darin Engelhardt, who supported this effort all the way.

I also thank the many friends who read bits of the manuscript as it was being written, including Peter Metz and John Ratey, both of whom appear in the memoir, and the many teachers who encouraged me to write and believe that life can turn out well.

Most of all I thank my family, which includes our twelve-year-old Jack Russell, Ziggy; my twenty-eight-year-old daughter, Lucy; my twenty-five-year-old son, Jack; and my twenty-two-year-old son, Tucker.

Which leaves my wife, Sue, whose age I won't reveal. Sue does it all for all of us. She radiates warmth and gives all she has every day to every person and every project she encounters. She is the kindest person I know. I could never thank her enough.

A NOTE ON THE AUTHOR

EDWARD M. HALLOWELL, MD, is the bestselling author of *Driven to Distraction* and many other acclaimed books, a leading authority in the field of ADHD, a world-renowned speaker, the host of the *Distraction* podcast, and the founder of the Hallowell Centers for Cognitive and Emotional Health in Boston MetroWest, New York City, San Francisco, and Seattle. He lives in Arlington, Massachusetts. www.drhallowell.com